TECHNICAL REFERENCE HANDBOOK

E. P. Rasis

AMERICAN TECHNICAL PUBLISHERS, INC.
ALSIP, ILLINOIS 60658

Library of Congress Catalog Number 83-73461
ISBN: 0-8269-3450-1

123456789-84-987654321

Printed in the United States of America

Art was provided courtesy of:

The American Society of Mechanical Engineers
Browning Manufacturing Division
Butterfield
Cincinnati Incorporated
Cincinnati Milacron
Clausing Division of Atlas Press Company
Cleveland Twist Drill Company
Deere and Company
Diamond Chain Company
Dietzgen Corporation
DoAll Company
Giddings and Lewis Machine Tool Co.
The L. S. Starrett Company
L. W. Chuck Co.
Machine Design
Mobil Oil Corporation
National Twist Drill-Division of Lear Siegler, Inc.
Norton Company
Regal-Beloit Corporation
South Bend Lathe, Inc.
Standard Tool Div. LSI
Sun Company

PREFACE

Technical Reference Handbook emphasizes the *fundamentals* of mathematics, machine tool technology, and physical science as they relate to industrial and technical applications. Particular attention was given in the selection of useful and relevant information developed throughout the years by manufacturers and technical associations. There are ample definitions provided to help insure comprehension of the various topics covered.

Over *300* practical formulas are included with examples of solutions to typical problems encountered in the classroom, the laboratory, and the shop. More than *240* illustrations, including charts and tables, support the text. The charts and tables are not a comprehensive reference source, but rather, are intended to show the kind of information that is available to individuals in industry and classroom settings.

The index at the back of the book is cross-referenced so that *key words* can be used to find information quickly. For example, if seeking information about cutting speeds for milling operations, page references will be found under *Cutting Speed* and under *Milling Machine*.

The Publishers

Contents

Chapter 1

MATHEMATICS

Mathematics is "the science of numbers and their operations, interrelations, combinations, generalizations, and abstractions." It is also the science of "space configurations and their structure, measurement, transformations, and generalizations." Mathematics includes *arithmetic, geometry, algebra,* and *trigonometry.*

ARITHMETIC

Arithmetic is the branch of mathematics that involves the computation [addition (+), subtraction (−), multiplication (×), and division (÷)] of positive *real numbers*. A real number is any number associated with a point on a line.

Fractions

A fraction is part of a whole unit or whole number. The top number of a fraction is the *numerator* and the bottom number is the *denominator*. The denominator of a fraction indicates the number of equal parts into which a unit is divided. The numerator indicates how many of these parts are used. For example, in the fraction $\frac{5}{16}$, the denominator 16 shows that the whole unit is divided into 16 equal parts, and the numerator 5 shows that 5 parts of the unit are used.

Proper Fractions. A proper fraction has a numerator smaller than its denominator. The value of a proper fraction is less than one unit.

Improper Fractions. An improper fraction has a numerator larger than its denominator. An improper fraction contains one, or more than one, whole unit. An improper fraction is changed into a *mixed number* (a whole number and a proper fraction) by dividing the numerator by the denominator. For example, when dividing the numerator 7 of the improper fraction $\frac{7}{2}$ by its denominator 2, the result is $3\frac{1}{2}$. The improper fraction $\frac{7}{2}$ contains three whole units and one half unit.

Developing and Reducing Fractions. To develop a fraction, multiply its numerator by any number, and multiply its denominator by that same number. To reduce a fraction, divide its numerator by any number, and divide the denominator by that same number. See Examples 1-1 and 1-2.

Adding and Subtracting Fractions. To add or subtract fractions, find the smallest denominator common to each fraction. This number is the *common*

Example 1-1: Develop the fraction $\frac{1}{2}$ into 16ths and 32nds.

Given: $\frac{1}{2}$ $\frac{\text{numerator}}{\text{denominator}}$

Solution: $\frac{1}{2} = \frac{1}{2} \times \frac{8}{8} = \frac{8}{16}$ and

$\frac{1}{2} \times \frac{16}{16} = \frac{16}{32}$

Example 1-2: Reduce the fraction $\frac{24}{32}$ into 8ths and 4ths.

Given: $\frac{24}{32}$

Solution: $\frac{24}{32} \div \frac{4}{4} = \frac{6}{8}$ and

$\frac{24}{32} \div \frac{8}{8} = \frac{3}{4}$

denominator. When adding fractions, add the numerators and write the sum over the common denominator. When subtracting fractions, subtract the smaller numerator from the larger numerator, and write the difference over the common denominator. See Examples 1-3 and 1-4.

NOTE: It is a good practice to reduce the resulting fraction into its lowest terms. If it is an improper fraction, change it into a mixed number.

Example 1-3: Add the fractions $\frac{1}{2}$, $\frac{3}{4}$, and $\frac{5}{16}$.

Given: $\frac{1}{2} + \frac{3}{4} + \frac{5}{16}$, and common denominator 16.

Solution: $\frac{1}{2} + \frac{3}{4} + \frac{5}{16} = \frac{1 \times 8}{2 \times 8} + \frac{3 \times 4}{4 \times 4} + \frac{5}{16}$

$= \frac{8}{16} + \frac{12}{16} + \frac{5}{16} = \frac{25}{16} = 1\frac{9}{16}$

Example 1-4: Subtract the fraction $\frac{3}{4}$ from $\frac{7}{8}$.

Given: $\frac{7}{8} - \frac{3}{4}$, and common denominator 8

Solution: $\frac{7}{8} - \frac{3}{4} = \frac{7}{8} - \frac{3 \times 2}{4 \times 2} = \frac{7}{8} - \frac{6}{8} = \frac{1}{8}$

Adding Fractions in Measurements. The addition of fractions is commonplace in machine shop work. This is because drawings of parts to be machined sometimes include only cumulative dimensions and not overall dimensions. Figure 1-1 shows the top view of a drill stand with fractional layout dimensions but without the overall length.

In Figure 1-1, six fractions that do not have a common denominator must be added. These fractions can be added only after finding a common denominator and converting them into fractions with that common denominator. See Example 1-5.

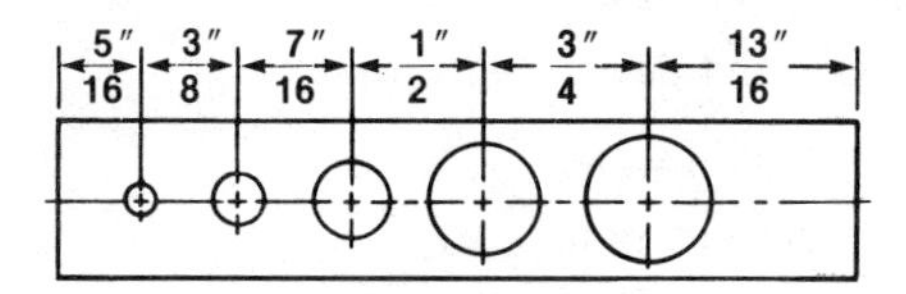

Figure 1-1. Top view of drill stand with fractional layout dimensions.

Example 1-5: Find the overall length of the drill stand (Figure 1-1).

Given: $\frac{5}{16} + \frac{3}{8} + \frac{7}{16} + \frac{1}{2} + \frac{3}{4} + \frac{13}{16}$, and common denominator 16.

Solution: $\frac{5}{16} + \frac{3 \times 2}{8 \times 2} + \frac{7}{16} + \frac{1 \times 8}{2 \times 8} + \frac{3 \times 4}{4 \times 4} + \frac{13}{16}$

$= \frac{5}{16} + \frac{6}{16} + \frac{7}{16} + \frac{8}{16} + \frac{12}{16} + \frac{13}{16} = \frac{51}{16} = 3\frac{3}{16}''$

Multiplying and Dividing Fractions. To multiply fractions, first multiply the numerators of each fraction by one another and place the product as the numerator of the new fraction. Then multiply the denominators by one another and place the product as the denominator of the new fraction.

To divide fractions, divide the numerator of the first fraction by the numerator of the second fraction, and place the quotient as the numerator of the new fraction. Then divide the denominator of the first fraction by the denominator of the second fraction, and place the quotient as the denominator of the new fraction.

Another method of dividing fractions is to invert the *divisor* (the number by which another number is divided), then proceed to multiply the fractions. See Examples 1-6 and 1-7.

NOTE: The numerator and denominator of a fraction can also be multiplied or divided by the same number (any number except zero) without changing the value of the fraction. For example, $\frac{5}{16} \times \frac{2}{2} = \frac{10}{32}$ or $\frac{10}{32} \div \frac{2}{2} = \frac{5}{16}$.

Example 1-6: Multiply the fraction $\frac{12}{16}$ by $\frac{1}{2}$.

Given: $\frac{12}{16} \times \frac{1}{2}$ or $\frac{3}{4} \times \frac{1}{2}$

Solution: $\frac{3}{4} \times \frac{1}{2} = \frac{3}{8}$

NOTE: It is always advisable to reduce fractions into their lowest terms prior to any computation. In this example, the fraction $\frac{12}{16}$ is reduced to $\frac{3}{4}$ by dividing its numerator and denominator by 4.

Example 1-7: Divide the fraction $\frac{3}{8}$ by $\frac{1}{4}$.

Given: $\frac{3}{8} \div \frac{1}{4}$

Solution: $\frac{3}{8} \div \frac{1}{4} = \frac{3}{2} = 1\frac{1}{2}$ or

$\frac{3}{8} \div \frac{1}{4} = \frac{3}{8} \times \frac{4}{1} = \frac{12}{8} = \frac{3}{2} = 1\frac{1}{2}$

Decimal Numbers

Decimal numbers are numbers that contain a *decimal point*. The decimal point is a period placed to the left of a number. Any number to the right of the decimal point is a fraction of a unit; that is, its value is less than one unit. For example, the value of the decimal number .4856, or 0.4856, is less than one unit. The first digit of the decimal number 0.4856 indicates four tenths (0.4) of one unit; the second digit indicates eight hundredths of one unit (0.08); the third digit indicates five thousandths of one unit (0.005); and the fourth digit indicates six ten thousandths of one unit (0.0006). The decimal number 0.4856 may be read as "point-four-eight-five-six," or as "four thousand, eight hundred, fifty-six ten thousandths."

Any decimal number can be expressed as a common fraction with the denominator 10, or some power of 10, such as 10^2 (100), 10^3 (1,000), 10^4 (10,000). For example, the decimal numbers 0.3, 0.45, 0.564, and 0.6879 can be expressed as the following common fractions:

$$0.3 = \frac{3}{10},\ 0.45 = \frac{45}{100},\ 0.564 = \frac{564}{1{,}000},\ \text{and } 0.6879 = \frac{6{,}879}{10{,}000}$$

NOTE: The addition of zeros to a number with a decimal point in front of it does not alter its value, but only changes its name.

SAME VALUES:

0.5	-	five tenths
0.50	-	fifty hundredths
0.500	-	five hundred thousandths
0.5000	-	five thousand ten thousandths

$$0.5 = 0.50 = 0.500 = 0.5000 \text{ or } \frac{5}{10} = \frac{50}{100} = \frac{500}{1{,}000} = \frac{5{,}000}{10{,}000}$$

Adding and Subtracting Decimal Numbers. To add or subtract decimal numbers, arrange the numbers so that the decimal points are aligned in a column. Proceed with the addition or subtraction as if the decimal numbers were whole numbers. To avoid errors, add zeros to the numbers with fewer digits beyond the decimal point. See Examples 1-8 and 1-9.

Example 1-8: Add the decimal numbers 4.16, 15.085, 125.500, 500, and 0.255.

Solution:

(a)		(b)	
4.16	addends	4.160	Preferred to avoid errors in addition
15.085		15.085	
125.500		125.500	
500.		500.000	
0.255		0.255	
645.000		645.000	

NOTE: The numbers to be added together are called *addends*.

Example 1-9: Subtract the decimal number 32.843 from 60.25.

Given: 60.25 − 32.843

Solution:

60.250	- minuend
− 32.843	- subtrahend
27.407	- difference or remainder

Multiplying Decimal Numbers. To multiply decimal numbers, proceed as if multiplying whole numbers. At the end of the computation, count the number of places after the decimal point of both the *multiplicand* (the number that is to be multiplied) and the *multiplier* (the number by which another number is multiplied). The number of places counted determines the position of the decimal point in the product. See Example 1-10.

Example 1-10: Multiply the decimal number 14.052 by 5.34.

Given: 14.052 × 5.34

Solution:

14.052	- multiplicand
× 5.34	- multiplier
56208	
42156	
70260	
75.03768	- product

Dividing Decimal Numbers. To divide decimal numbers, convert the *divisor* into a whole number by moving the decimal points of both the divisor and the *dividend* as many places to the right as needed to make the divisor a whole number. See Example 1-11.

Example 1-11: Divide the decimal number 35.075 by 2.5.

Given: $35.075 \div 2.5 = 350.75 \div 25$

Solution:

$$\begin{array}{rl} & \underline{\;\;14.03} \text{ - quotient} \\ \text{divisor - } 25) & 350.75 \text{ - dividend} \\ & \underline{25} \\ & 100 \\ & \underline{100} \\ & 00075 \\ & \underline{75} \\ & 00 \text{ - remainder} \end{array}$$

Changing Fractions into Decimal Numbers. To change a fraction into a decimal number, divide the numerator by its denominator. See Example 1-12.

Example 1-12: Change the fraction $\frac{3}{4}$ into a decimal number.

Given: $\frac{3}{4}$ or $3 \div 4$

Solution:

$$\begin{array}{rl} & \underline{\;\;.75} \\ 4) & 3.00 \\ & \underline{2\,8} \\ & 20 \\ & \underline{20} \\ & 00 \end{array}$$

Changing Decimal Numbers into Fractions. To change a decimal number into a fraction, multiply the decimal number by a fraction that has a denominator with the same numerical value as its numerator. See Example 1-13.

Changing fractions into decimals and decimals into fractions is essential when using measuring tools or instruments because some tools measure only in fractions of an inch, while others measure only in decimals. Whenever a decimal number cannot be changed exactly into one of the fractional parts of an inch, find the closest sixty-fourth. See Example 1-14.

Example 1-13: Change the decimal number 0.375 into two different fractions, one having a denominator of 4 and the other having a denominator of 8.

Given: $0.375 \times \frac{4}{4}$ and $0.375 \times \frac{8}{8}$

Solution: $0.375 \times \frac{4}{4} = \frac{1.500}{4} = \frac{1.5}{4}$

$0.375 \times \frac{8}{8} = \frac{3.000}{8} = \frac{3}{8}$

Example 1-14: Change the decimal number 0.582″ into the closest 64th of an inch.

Given: $0.582 \times \frac{64}{64}$

Solution: $0.582 \times \frac{64}{64} = \frac{37.248}{64} = \frac{37}{64}$

NOTE: The difference between $\frac{37.248}{64}$, or 0.582, and $\frac{37}{64}$, or 0.578, is only 0.004″. By using $\frac{37''}{64}$ instead of 0.582″, the measurement can be found with a steel rule graduated in 64ths.

Commonly used fractions associated with inch measurements are eighths, sixteenths, thirty-seconds, and sixty-fourths. Figure 1-2 shows the decimal equivalents of these fractions.

DECIMAL EQUIVALENTS of 8ths, 16ths, 32nds, 64ths

8ths	32nds	64ths	64ths
1/8 = .125	1/32 = .03125	1/64 = .015625	33/64 = .515625
1/4 = .250	3/32 = .09375	3/64 = .046875	35/64 = .546875
3/8 = .375	5/32 = .15625	5/64 = .078125	37/64 = .578125
1/2 = .500	7/32 = .21875	7/64 = .109375	39/64 = .609375
5/8 = .625	9/32 = .28125	9/64 = .140625	41/64 = .640625
3/4 = .750	11/32 = .34375	11/64 = .171875	43/64 = .671875
7/8 = .875	13/32 = .40625	13/64 = .203125	45/64 = .703125
16ths	15/32 = .46875	15/64 = .234375	47/64 = .734375
1/16 = .0625	17/32 = .53125	17/64 = .265625	49/64 = .765625
3/16 = .1875	19/32 = .59375	19/64 = .296875	51/64 = .796875
5/16 = .3125	21/32 = .65625	21/64 = .328125	53/64 = .828125
7/16 = .4375	23/32 = .71875	23/64 = .359375	55/64 = .859375
9/16 = .5625	25/32 = .78125	25/64 = .390625	57/64 = .890625
11/16 = .6875	27/32 = .84375	27/64 = .421875	59/64 = .921875
13/16 = .8125	29/32 = .90625	29/64 = .453125	61/64 = .953125
15/16 = .9375	31/32 = .96875	31/64 = .484375	63/64 = .984375

Figure 1-2. Decimal equivalent of common fractions.

Percentage, Ratio, and Proportion

Percentage is a fraction that has a denominator of 100, or a part of a whole expressed in hundredths. Percent (%) means "per hundred," or "out of each hundred." For example, 5% means five out of one hundred. Percentage can also be expressed as a decimal number by moving the decimal point two places to the left of the number. For example, 25% can be written as the fraction $^{25}/_{100}$, or the decimal number 0.25. To find the value of any percentage of a given number, multiply that number by the percentage. See Example 1-15. Figure 1-3 shows the fraction and decimal equivalents of selected percentages.

Example 1-15: Find 12% of $20,000.

Given: 12% of $20,000 or $\frac{12}{100} \times \$20{,}000$

Solution: $20{,}000 \times \frac{12}{100} = \frac{240{,}000}{100} = \$2{,}400$

PERCENT	FRACTION	DECIMAL	PERCENT	FRACTION	DECIMAL
10%	$^{1}/_{10}$	0.10	12.5%	$^{1}/_{8}$	0.125
20%	$^{2}/_{10}$	0.20	25%	$^{1}/_{4}$	0.250
30%	$^{3}/_{10}$	0.30	37.5%	$^{3}/_{8}$	0.375
40%	$^{4}/_{10}$	0.40	50%	$^{1}/_{2}$	0.500
50%	$^{5}/_{10}$	0.50	62.5%	$^{5}/_{8}$	0.625
60%	$^{6}/_{10}$	0.60	75%	$^{3}/_{4}$	0.750
70%	$^{7}/_{10}$	0.70	87.5%	$^{7}/_{8}$	0.875
80%	$^{8}/_{10}$	0.80			
90%	$^{9}/_{10}$	0.90			

Figure 1-3. Percentage equivalents.

A *ratio* is the quotient of two numbers of the same quantity indicating their relationship. For example, the ratio of one foot to one yard (three feet) is 1:3, or $^{1}/_{3}$. This ratio indicates that one yard is three times greater than one foot. The ratio of two numbers of the same quantity may be used to simplify computations, save time, and eliminate some of the common errors related to problems with metric and English units of measurement. For example, by using the ratio of one inch to one millimeter (1:25.4 or 0.03937) in a problem that contains both inches and millimeters, the problem is solved without making any unnecessary conversions. See Example 1-16.

Example 1-16: Find the area, in square inches, of a rectangular plane 1,016 mm long and 60″ wide.

Given: Length = 1,016 mm, Width = 60″
Area = Length × Width

Solution: *Convert millimeters (mm) to inches:* 1,016 × 0.03937 = 39.99992″

Find the area: 39.99992 × 60 = 2,399.9952 in^2

Solution without conversion: $\frac{1,016}{25.4} \times 60 = \frac{60,960}{25.4} = 2,400 \text{ in}^2$

A *proportion* is a mathematical expression indicating that two ratios are equal. For example, 1:2 = 4:8, or ½ = 4/8. If the numbers of a proportion are substituted with letters such as 1 = a, 2 = b, 4 = c, and 8 = d, the proportion may be written in any of the following forms:

$$\frac{1}{2} = \frac{4}{8} \text{ or } 1 \times 8 = 2 \times 4 \text{ or } 8 = 8 \quad \text{and}$$

$$\frac{a}{b} = \frac{c}{d} \text{ or } a \times d = b \times c$$

NOTE: In any proportion, the product of the numerator of the first fraction and the denominator of the second fraction is equal to the product of the denominator of the first fraction and the numerator of the second fraction.

GEOMETRY

Geometry is the branch of mathematics that involves problem-solving related to *plane figures* and *solid figures*.

Plane Figures

Plane figures (Figure 1-4) are characterized by two linear dimensions. The two dimensions are used to calculate area and perimeter.

The area, as well as the dimensions of a plane figure, can be found by applying a formula. Below are some of the letters and symbols used in plane figure formulas.

l or a = side (length) NOTE: a is also used to designate an altitude	A = area	r = radius of inscribed circle
w or b = side (width)	S = sum of sides (perimeter)	R = radius of circumscribed circle
h or c = side (height)	π = 3.1416	C = circumference
	d = diameter, or small diameter	D = diameter (larger of two circles)
	d′ = diagonal	

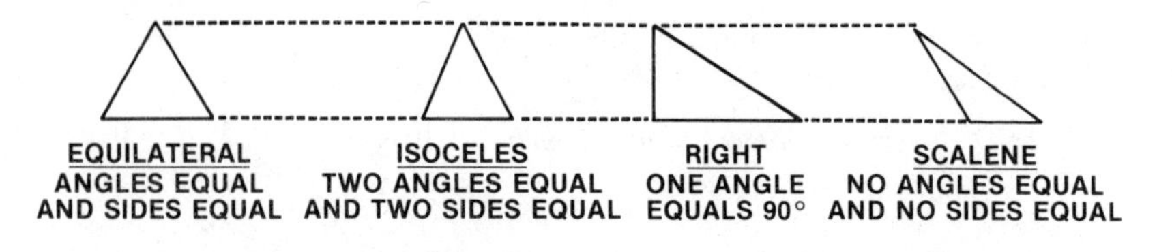

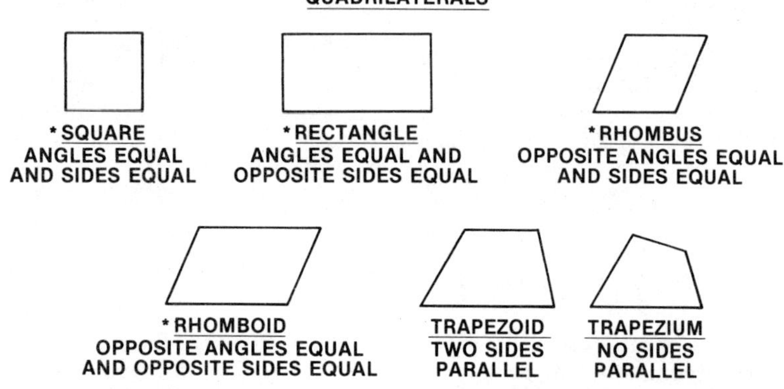

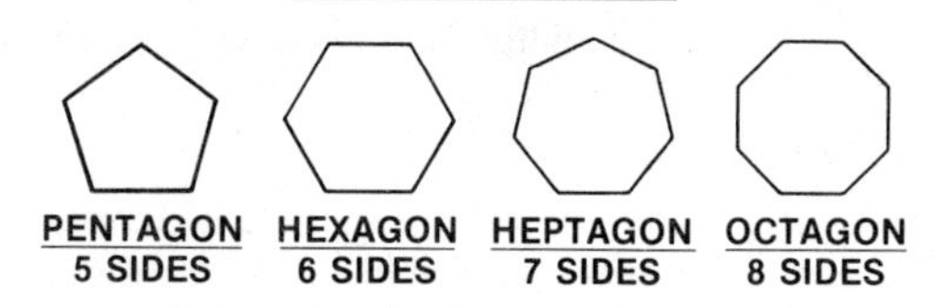

Figure 1-4. Plane (two dimensional) figures.

Common Plane Figures

1. The *triangle* (Figure 1-5) is a plane figure that has three sides (base, altitude, and hypotenuse) and three angles. The sum of these angles equals 180°. A triangle is formed by combining any of the following three angles (Figure 1-6): the *acute angle* (any angle smaller than 90°), the *right angle* (an angle of exactly 90°), and the *obtuse angle* (any angle greater than 90°). See Example 1-17A.

2. The *right triangle* (Figure 1-7) is a plane figure with three sides that form two acute angles and one right angle. See Example 1-17B.

3. A *square* (Figure 1-8) is a rectangular plane that has four equal sides and four right angles. See Example 1-17C.

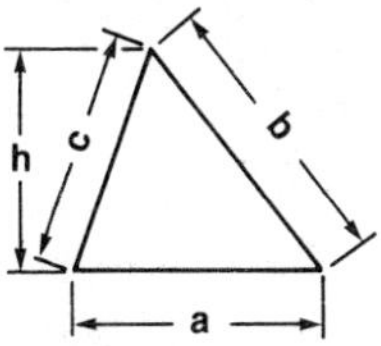

$$A = \frac{a \times h}{2}$$

$$a = \frac{2A}{h}, \quad h = \frac{2A}{a}$$

$$S = a + b + c$$

Figure 1-5. Triangle and related formulas.

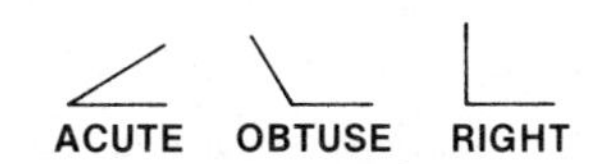

Figure 1-6. The three types of angles.

Example 1-17A: Find the area A of a triangle.
Given: Side a = 12″, height (altitude) = 14″

Solution: $A = \dfrac{a \times h}{2} = \dfrac{12 \times 14}{2} = \dfrac{168}{2} = 84 \text{ in}^2$

$$A = \frac{A \times b}{2}$$

$$c^2 = a^2 + b^2$$

$$a = \sqrt{c^2 + b^2}$$

$$b = \sqrt{c^2 - a^2}$$

$$c = \sqrt{a^2 + b^2}$$

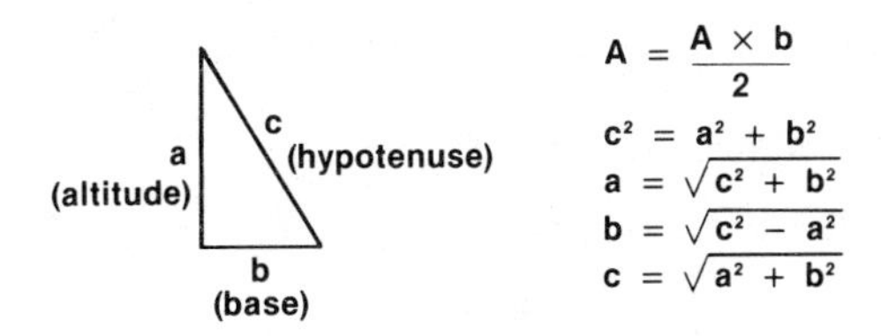

Figure 1-7. Right triangle and related formulas.

Example 1-17B: Find the hypotenuse c and the area A of a right triangle.
Given: Altitude a = 6″, Base b = 8″

Solution: $c = \sqrt{a^2 + b^2} = \sqrt{6^2 + 8^2} = \sqrt{36 + 64} = 10''$

$$A = \frac{a \times b}{2} = \frac{6 \times 8}{2} = \frac{48}{2} = 24 \text{ in}^2$$

$$A = a^2$$

$$d' = a\sqrt{2}$$

$$S = 4a$$

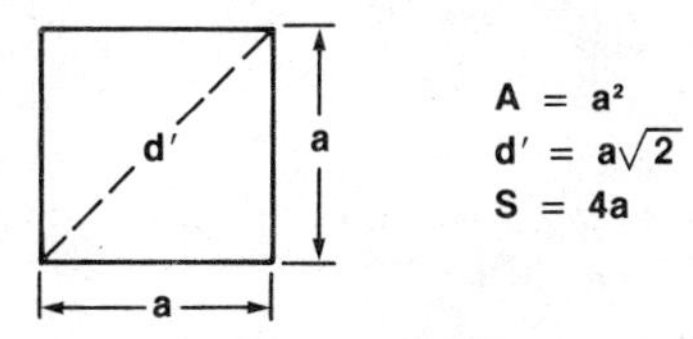

Figure 1-8. Square and related formulas.

Example 1-17C: Find the area A, the diagonal d′, and the sum of sides S of a square.
Given: Side a = 12″

Solution: $A = a^2 = 12^2 = 144 \text{ in}^2$

$d' = a\sqrt{2} = 12 \times 1.414 = 16.968''$

$S = 4 \times a = 4 \times 12 = 48''$

4. A *rectangle* (Figure 1-9) is a plane figure with two pairs of parallel and equal sides that form four right angles. See Example 1-17D.

5. A *rhombus* (Figure 1-10) is a plane figure with four equal sides that form two acute angles and two obtuse angles. See Example 1-17E.

6. A *trapezoid* (Figure 1-11) is a quadrilateral plane figure that has only two parallel sides. See Example 1-17F.

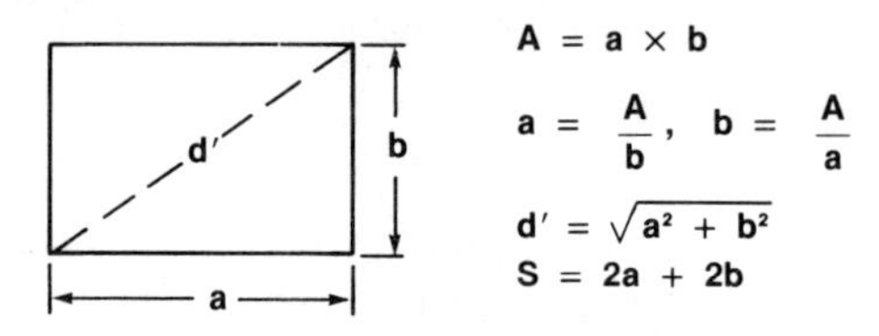

Figure 1-9. Rectangle and related formulas.

Example 1-17D: Find the side a, the diagonal d′, and the sum of the sides S of a rectangle.
Given: area A = 56 in², side b = 7″

Solution: $a = \frac{A}{b} = \frac{56}{7} = 8''$

$d' = \sqrt{a^2 + b^2} = \sqrt{8^2 + 7^2} = \sqrt{64 + 49} = \sqrt{113} = 10.63''$

$S = 2a + 2b = (2 \times 8) + (2 \times 7) = 16 + 14 = 30''$

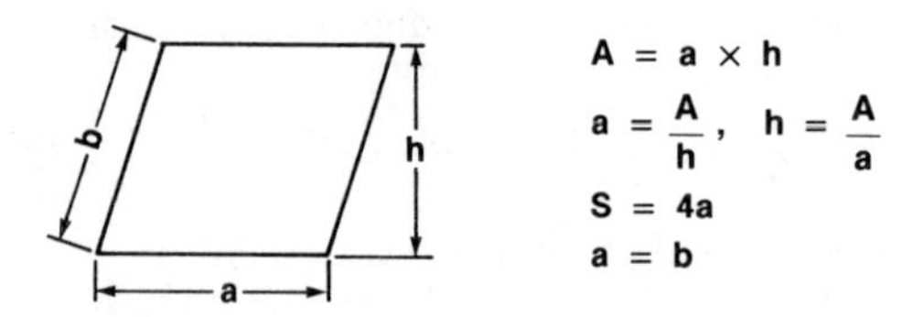

Figure 1-10. Rhombus and related formulas.

Example 1-17E: Find the area A, and the sum of sides S of a rhombus.
Given: Side a = 40″, height h = 32″

Solution: $A = a \times h = 40 \times 32 = 1{,}280 \text{ in}^2$

$S = 4 \times a = 4 \times 40 = 160''$

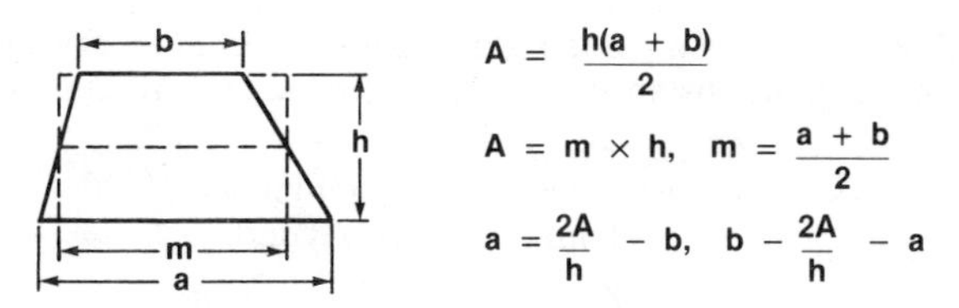

Figure 1-11. Trapezoid and related formulas.

Example 1-17F: Find the area A and the length m of a trapezoid.
Given: side a = 24″, side b = 20″, height h = 12″

Solution: $A = \dfrac{h(a + b)}{2} = \dfrac{12(24 + 20)}{2} = \dfrac{12 \times 44}{2} = \dfrac{528}{2} = 264 \text{ in}^2$

$m = \dfrac{a + b}{2} = \dfrac{24 + 20}{2} = \dfrac{44}{2} = 22''$

7. A *regular polygon* (Figure 1-12) is any plane figure with three or more sides that are equal in length. The diameter of a circumscribed circle D and an inscribed circle d for each regular polygon has a direct relationship with the length of its sides.

NOTE: The equilateral triangle and the square are also regular polygons.

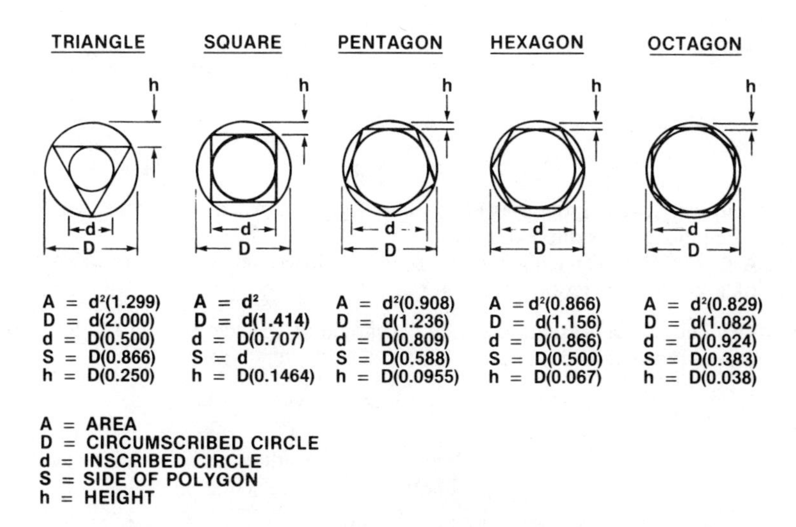

NOTE: HEIGHT (h) IS USED IN MILLING MACHINE OPERATIONS FOR SETTING PROPER DEPTH-OF-CUT WHEN MACHINING REGULAR POLYGON NUTS FROM ROUND STOCK.

Figure 1-12. Regular polygons and their formulas as related to circumscribed and inscribed circles about them.

8. A *circle* (Figure 1-13) is a plane figure enclosed within a curve. All points on a circle are of equal distance from the fixed center. All circles contain:

1. 360 degrees (360°)
   NOTE: 1 degree = 60 minutes
   1 minute = 60 seconds

2. 6.28 radians
   1 radian = 57° 17′ 46″ or 57.3°

See Example 1-17G.

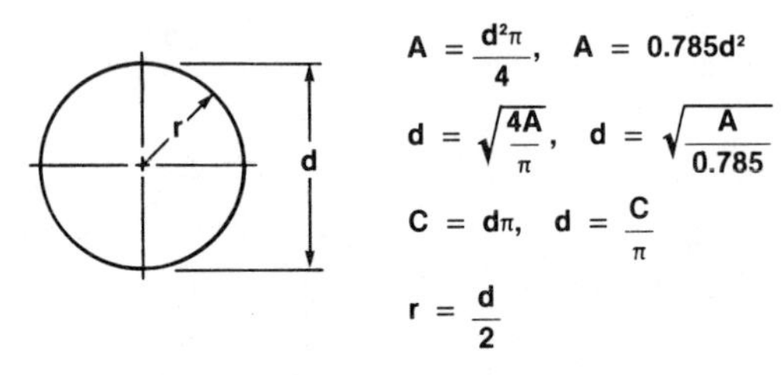

Figure 1-13. Circle and related formulas.

Example 1-17G: Find the area (A) and circumference (C) of a circle with a 20″ diameter.

Given: d = 20″

Solution: $A = \frac{d^2 \times \pi}{4} = \frac{20^2 \times 3.14}{4} = \frac{400 \times 3.14}{4} = \frac{1,256}{4} = 314 \text{ in}^2$ or

$A = 0.785 \times d^2 = 0.785 \times 20^2 = 0.785 \times 400 = 314.0 \text{ in}^2$

$C = d \times \pi = 20 \times 3.14 = 62.8''$

9. A *circular ring* (Figure 1-14) is a plane figure between two circles with the same center. See Example 1-17H.

10. A *sector* (Figure 1-15) is a plane figure enclosed between two radii and an arc of a circle. See Example 1-17I.

NOTE: Letter a represents the included angle of the sector, and the letter b represents the length of its arc.

d
D

$$A = \frac{D^2\pi}{4} - \frac{d^2\pi}{4}$$

$$A = \frac{\pi(D^2 - d^2)}{4}$$

$$D = \sqrt{\frac{4A}{\pi} + d^2}$$

$$d = \sqrt{D^2 - \frac{4A}{\pi}}$$

Figure 1-14. Circular ring and related formulas.

Example 1-17H: Find the area A of a circular ring.
Given: large diameter D = 12″, small diameter d = 10″

$$\text{Solution: } A = \frac{D^2 \times \pi}{4} - \frac{d^2 \times \pi}{4} = \frac{12^2 \times 3.14}{4} - \frac{10^2 \times 3.14}{4}$$

$$= \frac{144 \times 3.14}{4} - \frac{100 \times 3.14}{4} = \frac{452.16}{4} - \frac{314}{4}$$

$$= \frac{138.16}{4} = 34.54 \text{ in}^2$$

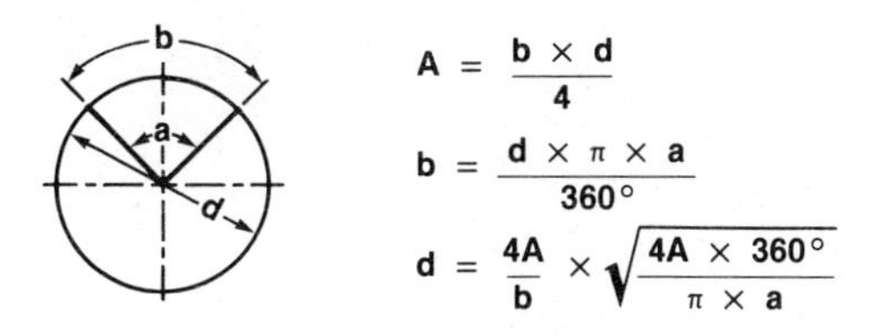

Figure 1-15. Sector and related formulas.

Example 1-17I: Find the length b and the area A of a sector.
Given: diameter of circle d = 8″, included angle a = 60°

$$\text{Solution: } b = \frac{d \times \pi \times a}{360°} = \frac{8 \times 3.14 \times 60°}{360°} = 25.12 \times 0.1666 = 4.19''$$

$$A = \frac{b \times d}{4} = \frac{4.19 \times 8}{4} = \frac{33.52}{4} = 8.38 \text{ in}^2$$

11. A *segment* (Figure 1-16) is a plane figure enclosed between the arc of a circle and the horizontal line (chord) which connects the two ends of the arc. See Example 1-17J.
12. An *ellipse* (Figure 1-17) is a closed curve plane figure characterized by a major diameter D and a minor diameter d. See Example 1-17K.

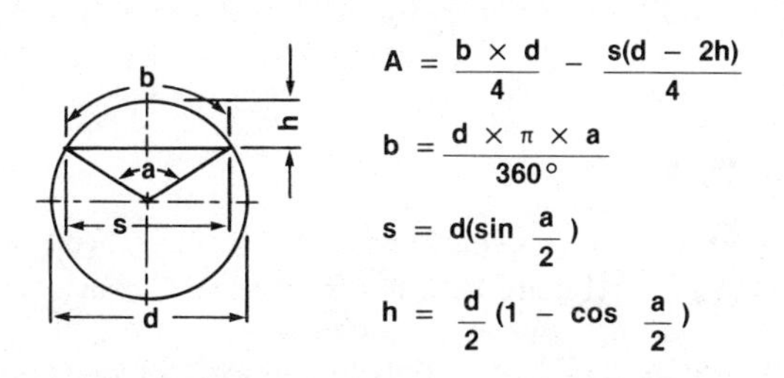

Figure 1-16. Segment and related formulas.

Example 1-17J: Find the length s and the height h of a segment.
Given: diameter of circle d = 10″, angle a = 60″,

$$\sin \frac{60}{2} = 0.500, \quad \cos \frac{60}{2} = 0.866$$

Solution: $s = d(\sin \frac{a}{2}) = 10 \times 0.500 = 5''$

$$h = \frac{d}{2}\left(1 - \cos \frac{a}{2}\right) = \frac{10}{2}\left(1 - 0.866\right) = 5 \times 0.134 = 0.670''$$

d

D

$$A = \frac{D \times d \times \pi}{4}$$

$$C = \frac{\pi(D + d)}{2}$$

$$C = 0.69 \times \sqrt{D^2 + d^2}$$

Figure 1-17. Ellipse and related formulas.

Example 1-17K: Find the area A and the circumference C of an ellipse.
Given: large diameter D = 10″, small diameter d = 6″

Solution: $A = \frac{D \times d \times \pi}{4} = \frac{10 \times 6 \times 3.14}{4} = \frac{188.4}{4} = 47.1 \text{ in}^2$

$$C = \frac{\pi(D + d)}{2} = \frac{3.14(10 + 6)}{2} = \frac{3.14 \times 16}{2} = 3.14 \times 8 = 25.12''$$

Solid Figures

Solid figures (Figure 1-18) are characterized by three linear dimensions. These dimensions are used to calculate volume and area. The symbols in the formulas used to determine dimensions of solid figures are the same as those used for plane figures. In addition, for solid figures, other symbols include:

V - volume, A_w - warped area, A_t - total area, s - side, D′ - diagonal

Common Solid Figures

1. A *cube* (Figure 1-19) has six equal and square surfaces. See Example 1-18A.
2. A *rectangular prism* (Figure 1-20) has six rectangular surfaces. See Example 1-18B.
3. A *pyramid* (Figure 1-21) has a polygonal base and triangular sides that meet in a common *vertex*. See Example 1-18C.

PRISMS

RIGHT SQUARE   RIGHT RECTANGULAR   OBLIQUE RECTANGULAR

PYRAMIDS

VERTEX
BASE
ALTITUDE

RIGHT TRIANGULAR   RIGHT SQUARE (Truncated)   OBLIQUE PENTAGONAL

CONES

VERTEX
AXIS
ALTITUDE
BASE

RIGHT CIRCULAR   OBLIQUE CIR (Frustum)   OBLIQUE CIR (Truncated)

CYLINDERS

BASE
BASE
AXIS
ALTITUDE
AXIS
POLE

RIGHT CIRCULAR   OBLIQUE CIRCULAR   SPHERE

Figure 1-18. Solid (three dimensional) figures.

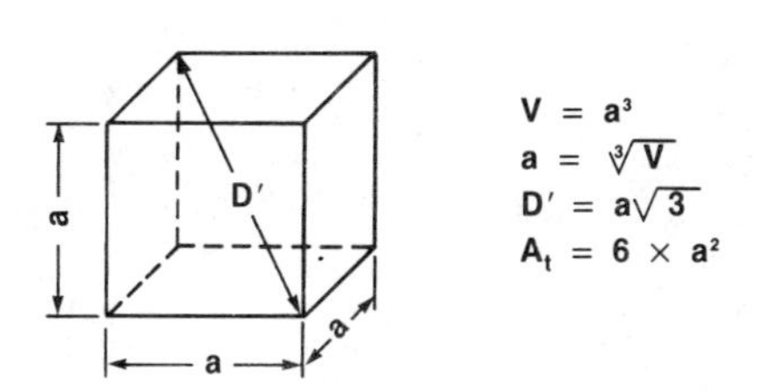

Figure 1-19. Cube and related formulas.

Example 1-18A: Find the volume V, the diagonal D′, and the total area A_t of a cube.

Given: Side a = 8″

Solution: $V = a^3 = 8^3 = 512 \text{ in}^3$

$D' = a\sqrt{3} = 8 \times 1.732 = 13.85''$

$A_t = 6 \times a^2 = 6 \times 8^2 = 6 \times 64 = 384 \text{ in}^2$

4. A *frustum pyramid* (Figure 1-22) is a pyramid-shaped solid figure formed by cutting off its top by a plane parallel to its base. See Example 1-18D.

5. A *cone* (Figure 1-23) is generated by the rotation of a right triangle about its altitude. The altitude is the axis of the cone and it corresponds to the height of the cone. The slant length corresponds to the hypotenuse of the right triangle. See Example 1-18E.

$$V = a \times b \times h$$

$$a = \frac{V}{b \times h}, \quad h = \frac{V}{a \times b}$$

$$D' = \sqrt{a^2 + b^2 + h^2}$$

$$A_t = 2(ab + bh + ah)$$

Figure 1-20. Rectangular prism and related formulas.

Example 1-18B: Find the volume V, the diagonal D′, and the total area A_t of a rectangular prism.

Given: Side a = 18″, Side b = 8″, Side h = 6″

Solution: $V = a \times b \times h = 18 \times 8 \times 6 = 144 \times 6 = 864 \text{ in}^3$

$$D' = \sqrt{a^2 + b^2 + h^2} = \sqrt{324 + 64 + 36} = \sqrt{424} = 20.59''$$

$$A_t = 2(ab + bh + ah) = 2(144 + 48 + 108) = 2(300) = 600 \text{ in}^2$$

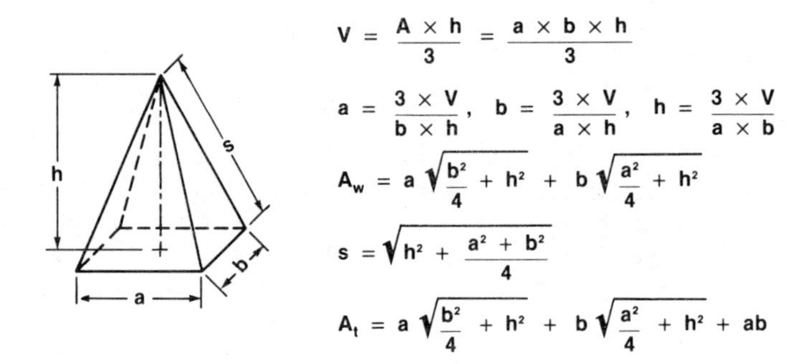

Figure 1-21. Pyramid and related formulas.

6. A *frustum cone* (Figure 1-24) is a cone-shaped solid figure formed by cutting off its top by a plane parallel to its base. See Example 1-18F.

7. A *cylinder* (Figure 1-25) is composed of two circles at opposite ends of each other for bases, and a curved surface perpindicular to the bases. A cylinder is generated by the rotation of a rectangle about a side as an axis. See Example 1-18G.

8. A *hollow cylinder* (Figure 1-26) is a cylinder with a hollow center. The axis of the hole coincides with the axis of the cylinder. See Example 1-18H.

9. A *sphere* (Figure 1-27) is a solid figure that has a curved surface. Every point of this surface is of equal distance from the center point of the sphere. See Example 1-18I.

10. A *spherical segment* (Figure 1-28) is formed by passing a plane through a sphere. If the plane passes through the center of the sphere, it is equal to one half the size of the sphere. See Example 1-18J.

Example 1-18C: Find the volume V, side s, warped area A_w, and total area A_t of a pyramid.
Given: side a = 6″, side b = 4″, height h = 10″

Solution: $V = \frac{a \times b \times h}{3} = \frac{6 \times 4 \times 10}{3} = \frac{240}{3} = 80 \text{ in}^3$

$$s = \sqrt{h^2 + \frac{a^2 + b^2}{4}} = \sqrt{10^2 + \frac{6^2 + 4^2}{4}} = \sqrt{100 + \frac{36 + 16}{4}}$$

$$= \sqrt{100 + \frac{52}{4}} = \sqrt{100 + 13} = \sqrt{113} = 10.630''$$

$$A_w = a\sqrt{\frac{b^2}{4} + h^2} + b\sqrt{\frac{a^2}{4} + h^2}$$

$$= 6\sqrt{\frac{(4)^2}{4} + 10^2} + 4\sqrt{\frac{(6)^2}{4} + 10^2}$$

$$= 6\sqrt{\frac{16}{4} + 100} + 4\sqrt{\frac{36}{4} + 100}$$

$$= 6\sqrt{104} + 4\sqrt{109} = 6(10.198) + 4(10.440)$$

$$= 61.19 + 41.76 = 102.95 \text{ in}^2$$

$$A_t = a\sqrt{\frac{b^2}{4} + h^2} + b\sqrt{\frac{a^2}{4} + h^2} + a \times b$$

$$= 6\sqrt{\frac{(4)^2}{4} + 10^2} + 4\sqrt{\frac{(6)^2}{4} + 10^2} + 6 \times 4$$

$$= 6\sqrt{\frac{16}{4} + 100} + 4\sqrt{\frac{36}{4} + 100} + 24$$

$$= 6\sqrt{104} + 4\sqrt{109} + 24 = 6(10.198) + 4(10.440) + 24$$

$$= 61.19 + 41.76 + 24 = 126.95 \text{ in}^2$$

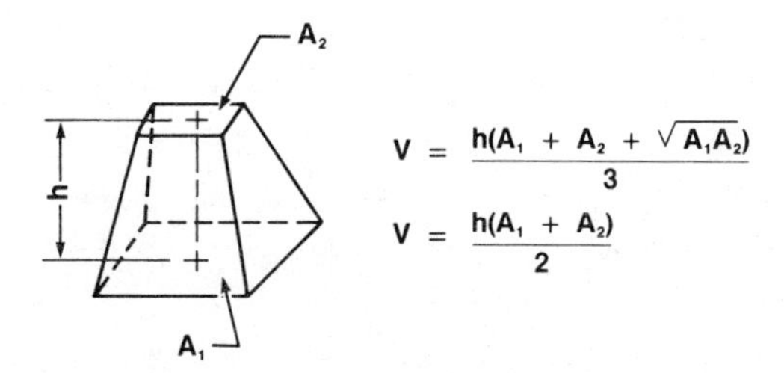

Figure 1-22. Frustum pyramid and related formulas.

Example 1-18D: Find the volume V of a frustum pyramid.
Given: area $A_1 = 30\ \text{in}^2$, area $A_2 = 20\ \text{in}^2$, height $h = 9''$

Solution: $$V = \frac{h(A_1 + A_2 + \sqrt{A_1 \times A_2})}{3}$$

$$= \frac{9(30 + 20 + \sqrt{30 \times 20})}{3} = 3(50 + 24.49) = 223.47\ \text{in}^3$$

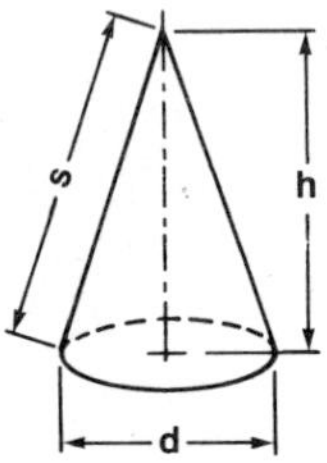

$$V = \frac{A \times h}{3} = \frac{d^2 \times \pi \times h}{12}$$

$$V = \frac{A \times h}{3} = \frac{d^2 \times \pi \times h}{12}$$

$$s = \sqrt{h^2 + \frac{d^2}{4}}$$

$$A_w = \frac{d \times \pi \times s}{2}$$

Figure 1-23. Cone and related formulas.

Example 1-18E: Find the volume V, side s, and warped area A_w of a cone.
Given: diameter $d = 18''$, height $h = 24''$

Solution: $$V = \frac{d^2 \times \pi}{4} \times \frac{h}{3} = \frac{18^2 \times 3.14}{4} \times \frac{24}{3}$$

$$= 254.34 \times 8 = 2{,}034.72\ \text{in}^3$$

$$s = \sqrt{h^2 + \frac{d^2}{4}} = \sqrt{24^2 + \frac{18^2}{4}} = \sqrt{576 + 81} = 25.63''$$

$$A_w = \frac{d \times \pi \times s}{2} = \frac{18 \times 3.14 \times 25.63}{2}$$

$$= \frac{1{,}448.6}{2} = 724.30\ \text{in}^2$$

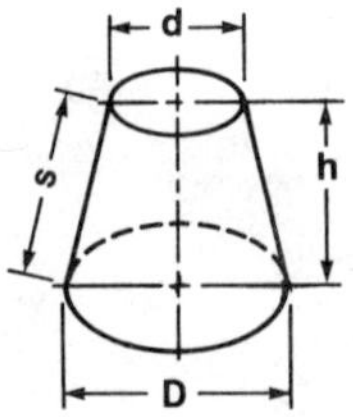

$$V = \frac{\pi \times h(D^2 + d^2 + Dd)}{12}$$

$$V = \frac{(A_1 + A_2)h}{2}$$

$$s = \frac{\sqrt{4h^2 + (D - d)^2}}{2}$$

$$A_w = \frac{\pi \times s(D + d)}{2}$$

Figure 1-24. Frustum cone and related formulas.

Example 1-18F: Find the volume V, side s, and warped area A_w of a frustum cone.
Given: large diameter D = 20″, small diameter d = 15″, h = 50″

Solution: $V = \frac{\pi \times h}{12}(D^2 + d^2 + D \times d) \frac{3.14 \times 50}{12}(20^2 + 15^2 + 20 \times 15)$

$$= \frac{157}{12}(400 + 225 + 300) = 947.428 \text{ in}^3$$

$$s = \frac{1}{2}\sqrt{4h + (D - d)^2} = \frac{1}{2}\sqrt{200 + 5^2} = \frac{1}{2}\sqrt{225}$$

$$= \frac{1}{2} \times 15 = 7.5''$$

$$A_w = \frac{\pi \times s}{2}(D + d) = \frac{3.14 \times 7.5}{2}(20 + 15) = 11.775 \times 35$$

$$= 412.125 \text{ in}^2$$

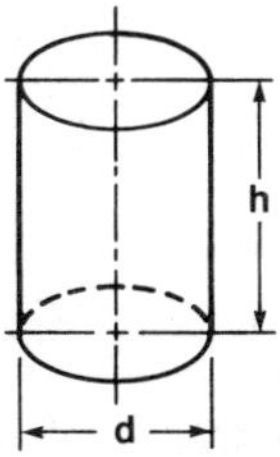

$V = A \times h = \frac{d^2 \times \pi \times h}{4}$

$d = \sqrt{\frac{4 \times V}{\pi \times h}}, \quad h = \frac{V}{A}$

$A_w = \pi \times d \times h$

$A_t = \pi \times d \times h + \frac{d^2}{2}$

Figure 1-25. Cylinder and related formulas.

Example 1-18G: Find the volume V, the warped area A_w, and the total area A_t of a cylinder.
Given: diameter d = 20″, height h = 90″

Solution: $V = \frac{d^2 \times \pi}{4} \times h = \frac{400 \times 3.14}{4} \times 90 = 314 \times 90$

$$= 28{,}260 \text{ in}^3$$

$$A_w = \pi \times d \times h = 3.14 \times 20 \times 90 = 5{,}652 \text{ in}^2$$

$$A_t = \pi \times d\left(h + \frac{d}{2}\right) = 3.14 \times 20\left(90 + \frac{20}{2}\right) = 62.8 \times 100$$

$$= 6{,}280 \text{ in}^2$$

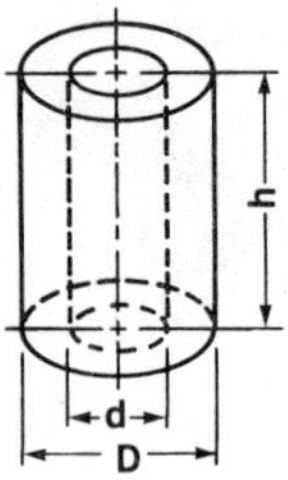

$$V = \frac{\pi \times h(D^2 - d^2)}{4}$$

$$h = \frac{4 \times V}{\pi(D^2 - d^2)}$$

$$D = \sqrt{\frac{4 \times V}{\pi \times h} + d^2}$$

$$d = \sqrt{D^2 - \frac{4 \times V}{\pi \times h}}$$

Figure 1-26. Hollow cylinder and related formulas.

Example 1-18H: Find the volume V of a hollow cylinder.
Given: diameter D = 18″, diameter d = 16″, height h = 54″

Solution: $V = \frac{\pi \times h}{4}(D^2 - d^2) = \frac{3.14 \times 54}{4}(324 - 256) = 42.39 \times 68$

$= 2{,}882.52 \text{ in}^3$

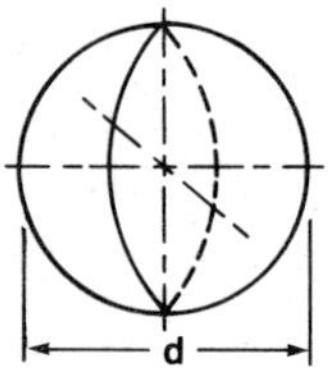

$$V = \frac{d^3 \times \pi}{6} \quad \text{or} \quad V = 0.523d^3$$

$$d = \sqrt[3]{\frac{6 \times V}{\pi}}$$

$$A_t = d^2 \times \pi$$

Figure 1-27. Sphere and related formulas.

Example 1-18I: Find the volume V and the total area A_t of a sphere.
Given: diameter d = 12″

Solution: $V = \frac{d^3 \times \pi}{6} = \frac{12^3 \times 3.14}{6} = \frac{1{,}728 \times 3.14}{6} = 904.32 \text{ in}^3$

$A_t = d^2 \times \pi = 12^2 \times 3.14 = 144 \times 3.14 = 452.16 \text{ in}^2$

$$V = \frac{\pi \times h^2(3d - 2h)}{6}$$

$$A_w = \pi \times d \times h$$

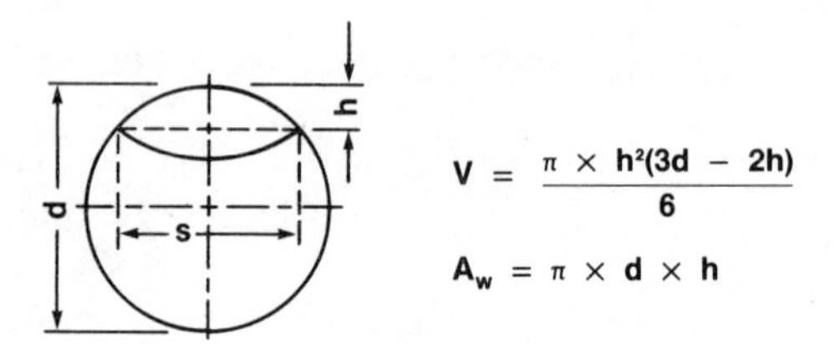

Figure 1-28. Spherical segment and related formulas.

Example 1-18J: Find the volume V and the warped area A_w of a spherical segment.
Given: diameter d = 15″, height h = 5″

Solution: $V = \frac{\pi \times h^2}{6}(3d - 2h) = \frac{3.14 \times 5^2}{6}(45 - 10) = 13.08 \times 35$

$= 457.8 \text{ in}^3$

$A_w = d \times \pi \times h = 15 \times 3.14 \times 5 = 235.5 \text{ in}^2$

ALGEBRA

Algebra is the branch of mathematics that studies arithmetic through the use of symbols and both positive (+) and negative (−) numbers. Negative numbers are the points on a straight line located to the left of the zero point. Positive numbers are the points on a straight line located to the right of the zero point.

−4 −3 −2 −1 0 1 2 3 4
NEGATIVE NUMBERS ⟵ 0 ⟶ POSITIVE NUMBERS

Laws and Properties of Real Numbers

All the rules of arithmetic are applicable to algebra; thus, the substitution of numbers with letters makes difficult arithmetic operations simpler. If letters are assigned to values, such as a = 1, b = 2, c = 3, and d = 4, then the laws of real numbers can be stated in both arithmetical and algebraic expressions.

Closure Law. The Closure Law (Figure 1-29) states that if a and b are real numbers, then their sum (a + b) and their product (a × b) are unique real numbers.

Associative Law. The Associative Law (Figure 1-29) states that in addition and multiplication, the terms can be associated (paired or grouped together) in any manner without affecting the sum or product.

Identity Law. The Identity Law (Figure 1-29) states that the sum of any number and zero is equal to that number, and that the product of any number and one is equal to that number.

Inverse Law. The Inverse Law (Figure 1-29) states that the sum of two identical numbers with opposite signs (one positive and one negative) is equal to zero. The product of a number and its inverse is equal to one.

Commutative Law. The Commutative Law states that the terms in a sum, or the factors in a product, can be commuted (positions of terms can be interchanged) without affecting the sum or product.

Distributive Law. The Distributive Law states that the terms within a factor can be added first and then multiplied by the cofactor, or each term can be multiplied by the cofactor first and then added to their products.

LAWS OF REAL NUMBERS	ARITHMETICAL EXPRESSIONS	ALGEBRAIC EXPRESSIONS
Addition:		
1. Closure Law	$2 + 3 = 5$	a + b is a unique real number
2. Associative Law	$(2 + 3) + 4 = 2 + (3 + 4)$	$(a + b) + c = a + (b + c)$
3. Identity Law	$2 + 0 = 0 + 2 = 2$	$a + 0 = 0 + a = a$
4. Inverse Law	$2 + (-2) = (-2) + 2 = 0$	$a + (-a) = (-a) + a = 0$
5. Commutative Law	$2 + 3 = 3 + 2$	$a + b = b + a$
Multiplication:		
1. Closure Law	$2 \times 3 = 6$	a × b is a unique real number
2. Associative Law	$(2 \times 3) \times 4 = 2 \times (3 \times 4)$	$(a \times b) \times c = a \times (b \times c)$
3. Identity Law	$2 \times 1 = 1 \times 2 = 2$	$a \times 1 = 1 \times a = a$
4. Inverse Law	$2 \times 1/2 = 1/2 \times 2 = 1$	$a \times 1/a = 1/a \times a = 1$
5. Commutative Law	$2 \times 3 = 3 \times 2$	$a \times b = b \times a$
6. Distributive Law	$2 \times (3 + 4) = (2 \times 3) + (2 \times 4)$	$a \times (b + c) = (a \times b) + (a \times c)$

Figure 1-29. Laws of real numbers and their arithmetical and algebraic expressions.

Rules for Algebraic Expressions

Four rules apply in solving algebraic problems.

Rule 1: Adding numbers with like signs results in a sum with that same sign.

$$(+2a) + (+5a) = (+7a) = 7a$$
$$(-2a) + (-5a) = (-7a) = -7a$$

Rule 2: Adding numbers with unlike signs results in a sum with the sign of the number with the larger absolute value.

$$(-2a) + (+8a) = (+6a) = 6a$$
$$(+2a) + (-8a) = (-6a) = -6a$$

Rule 3: The product of two numbers with like signs is always a positive number.

$$(-2a) \times (-2) = -(-4a) = 4a$$

Rule 4: The product of two numbers with unlike signs is always a negative number.

$$-2 \times (+2a) = -4a$$

Grouping Symbols

The grouping symbols that are used in algebraic expressions are the *parenthesis* ( ), the *bracket* [ ], and the *brace* { }. These symbols are used whenever several operations are to be performed, and to indicate the order of these operations. Omission of a grouping symbol may change the results of calculations. For example, in the problem

$$(a + b) \times (x + y + z)$$

omission of the first pair of parentheses would result in

$$a + b \times (x + y + z)$$

which alters the final answer. See Examples 1-19 and 1-20.

Example 1-19: Multiply the algebraic expression $(a + b)$ by $(x + y + z)$.
Given: $(a + b) \times (x + y + z)$
Solution: $(a + b) \times (x + y + z)$
$= ax + ay + az + bx + by + bz$

Example 1-20: Multiply the algebraic expression $a + b$ by $(x + y + z)$.
Given: $a + b \times (x + y + z)$
Solution: $a + b \times (x + y + z)$
$= a + bx + by + bz$

NOTE: Terms or factors within grouping symbols constitute a part that must be treated as a single unit.

Removing Grouping Symbols

Grouping symbols can be removed by multiplying each enclosed term by 1, or by -1 if there is a minus sign in front of the symbol. See Examples 1-21 and 1-22.

Example 1-21: Remove the parentheses from the algebraic expression $a - (b - c)$.
Given: $a - (b - c)$
Solution: $a - (b - c) = a - b + c$

Example 1-22: Remove the parentheses from the algebraic expression $a + (b - c)$.
Given: $a + (b - c)$
Solution: $a + (b - c) = a + b - c$

In complex algebraic expressions containing more than one type of grouping symbol, an ordered procedure must be followed for their removal. First, remove the parentheses; second, remove the brackets; third, remove the braces. After the grouping symbols are removed, proceed with the addition of similar factors. See Example 1-23.

Example 1-23: Remove the grouping symbols from the algebraic expression

$$2\{3[a + b(a + 4)] + a(b + 2)\}$$

Solution: *Removal of parentheses:* $2\{3[a + ba + 4b] + ab + 2a\}$

Removal of brackets: $= 2\{3a + 3ba + 12b + ab + 2a\}$

Removal of braces: $= 6a + 6ba + 24b + 2ab + 4a$

Addition of similar factors: $= 10a + 8ab + 24b$

EQUATIONS

An equation is a statement of equality between two algebraic expressions. In every equation, the numerical value on the left side of the equal sign is equal to the numerical value on the right side of the equal sign. For example, in the equation

$$a \times b = c$$

if the letters are substituted with numbers, such as $a = 2$, $b = 4$, and $c = 8$, then $a \times b = c$ would become

$$2 \times 4 = 8$$

Using Axioms to Solve Equations

An axiom is a self-evident principle or rule of general acceptance. Four axioms are used to facilitate the solution of equations.

First Axiom. The first axiom states that if equals are added to equals, the sums are equal. For example, in the equation

$$a + b = c$$

the value (b) can be added to each side of the equation:

$$a + b + (b) = c + (b) \quad \text{or}$$
$$a + 2b = c + b.$$

Second Axiom. The second axiom states that if equals are subtracted from equals, the remainder is equal. For example, in the equation

$$a + b = c$$

the value (b) can be subtracted from each side of the equation:

$$a + b - (b) = c - (b), \quad \text{or}$$
$$a = c - b.$$

NOTE: Any term of an equation may be transposed (moved) from one side of the equation to the other by changing its sign. For example, to transpose the b in the equation $a + b = c$, change its sign to negative and move it to the other side of the equal sign. The result is $a = c - b$.

Third Axiom. The third axiom states that if equals are multiplied by equals, the products are equal. For example, in the equation

$$\frac{a}{b} = c$$

the value (b) can be multiplied to each side of the equation:

$$\frac{a \times (b)}{b} = c \times (b) \text{ or } \quad a = bc$$

Fourth Axiom. The fourth axiom states that if equals are divided by equals, the quotients are equal. For example, in the equation

$$a \times b = b \times c,$$

each side of the equation can be divided by the value (b), resulting in

$$\frac{a \times b}{(b)} = \frac{b \times c}{(b)} \quad \text{or} \quad a = c$$

Solving Equations

The solution of any equation is found in three steps:
Step 1: Remove any grouping symbols, if applicable.
Step 2: Group all the factors of the unknown quantity in the form of a product or quotient on the left side of the equal sign, and group the known quantity on the right side of the equal sign. This is accomplished by using the first and second axioms.
Step 3: Isolate the unknown quantity on the left side of the equation by using the third or fourth axioms. See Examples 1-24 and 1-25.

Example 1-24: Solve the equation $4(x - 2) = 2x + 2$

Given: $4(x - 2) = 2x + 2$

Solution: *Step 1—remove grouping symbols* $\quad 4x - 8 = 2x + 2$

Step 2—first and second axioms applied $\quad 4x - 2x = 2 + 8$

$$2x = 10$$

Step 3—third and fourth axioms applied

$$x = \frac{10}{2}$$

$$x = 5$$

Example 1-25: Solve the equation $a^2 + 2a + 20 = a^2 + 60$

Given: $a^2 + 2a + 20 = a^2 + 60$

Solution: *Step 1 unnecessary because there are no grouping symbols.*

Step 2—first and second axioms applied. $a^2 - a^2 + 2a = 60 - 20$

$2a = 40$

Step 3—third and fourth axioms applied. $a = \frac{40}{2}$

$a = 20$

MATHEMATICAL FORMULAS

A mathematical formula is an equation indicating the relationship between the quantities on the left side of the equal sign and the quantities on the right side of the equal sign. Using formulas to solve technical problems may present some difficulties only if the unknown quantity is on the right side of the equation. These difficulties are eliminated by applying the four axioms for the solution of equations.

Problem-solving Using Formulas

The formula for calculating the volume of a rectangular solid figure is:

$V = l \times w \times h$

V = volume
l = length
w = width
h = height

The various factors of a formula can be arranged so that the unknown quantity is isolated on the left side. See Example 1-26.

Example 1-26: Use the formula $V = l \times w \times h$ to determine l, w, and h.

Given: $V = l \times w \times h$ or $l \times w \times h = V$

Solution: (1) $l \times w \times h = V$ or $\frac{l \times w \times h}{w \times h} = \frac{V}{w \times h}$ and

$l = \frac{V}{w \times h}$

(2) $w \times l \times h = V$ or $\frac{w \times l \times h}{l \times h} = \frac{V}{l \times h}$ and

$w = \frac{V}{l \times h}$

(3) $h \times w \times l = V$ or $\frac{h \times w \times l}{w \times l} = \frac{V}{w \times l}$ and

$h = \frac{V}{w \times l}$

The formula for calculating voltage in terms of watts and amperes is:

$$E = \frac{P}{I}$$

E = volts
P = watts
I = amperes

Example 1-27 shows an application of this formula.

Example 1-27: Use the formula $E = \frac{P}{I}$ to determine P and I.

Given: $E = \frac{P}{I}$ or $\frac{P}{I} = E$

Solution: (1) $\frac{P}{I} = E$ or $\frac{P \times I}{I} = E \times I$ and $P = E \times I$

(2) $E \times I = P$ or $\frac{E \times I}{E} = \frac{P}{E}$ and $I = \frac{P}{E}$

The formula for determining the taper-per-inch of a tapered machine element is:

$$Tpi = \frac{D - d}{l}$$

Tpi = taper-per-inch
D = large diameter
d = small diameter
l = length

Example 1-28 shows an application of this formula.

Example 1-28: Use the formula $Tpi = \frac{D - d}{l}$ to determine D, d, and l

Given: $Tpi = \frac{D - d}{l}$ or $\frac{D - d}{l} = Tpi$

Solution: $\frac{D - d}{l} = Tpi$ or $\frac{(D - d)l}{l} = Tpi \times l$

$D - d = Tpi \times l$ and

(1) $D = Tpi \times l + d$

(2) $d = D - Tpi \times l$

(3) $l = \frac{D - d}{Tpi}$

ROOTS

The root of a number is a quantity that, when multiplied by itself a given number of times, results in the original number. For example, the square root of 25, expressed as $\sqrt{25}$, is equal to 5 because $5 \times 5 = 25$. Another example is the fifth root of 243, expressed as $\sqrt[5]{243}$. It is equal to 3 because $3 \times 3 \times 3 \times 3 \times 3 = 243$.

Multiplying Roots

Multiplying roots can be done by either extracting the root of each term then multiplying the results, or by treating the product of the terms as if they were under the same *radical sign* ($\sqrt{\ }$), then extracting the root. See Example 1-29.

Example 1-29: Multiply the square root of 9 by the square root of 4.

Solution: $\sqrt{9} \times \sqrt{4} = 3 \times 2 = 6$ or
$\sqrt{9 \times 4} = \sqrt{36} = 6$

Dividing Roots

The principles used for multiplying roots is also used for dividing roots. See Example 1-30.

Example 1-30: Divide the square root of 64 by the square root of 16.

Solution: $\dfrac{\sqrt{64}}{\sqrt{16}} = \dfrac{8}{4} = 2$ or $\sqrt{\dfrac{64}{16}} = \sqrt{4} = 2$

Factoring Roots

Factoring is the process of analyzing a number by separating its component parts (factors) and presenting them as a product with the same numerical value. For example, the number 24 can be analyzed and presented as the product of the following factors:

$$24 = 2 \times 2 \times 2 \times 3$$
$$24 = 2 \times 3 \times 4$$
$$24 = 3 \times 8$$
$$24 = 4 \times 6$$

Factoring terms under a radical sign is done by following the reverse procedure that is used for multiplying roots. See Example 1-31.

NOTE: The square of any root is equal to the number within the radical sign. For example, the square of the square root of $(\sqrt{4})^2 = 4$. This principle is used to remove roots from the denominator of fractions in order to avoid complicated calculations. See Figure 1-30 for the formulas associated with roots.

Example 1-31: Factor the square root of 32.

Given: $\sqrt{32}$ - Formula $\sqrt{c} = \sqrt{a^2} \times \sqrt{b}$

Solution: $\sqrt{32} = \sqrt{4^2} \times \sqrt{2} = \sqrt{16} \times \sqrt{2}$

$x\sqrt[n]{a} \pm y\sqrt[n]{a} = x \pm y\sqrt[n]{a}$	$2\sqrt[3]{a} + 5\sqrt[3]{a} = 7\sqrt[3]{a}$
$\sqrt[n]{a \times b} = \sqrt[n]{a} \times \sqrt[n]{b}$	$\sqrt[4]{16 \times 81} = \sqrt[4]{16} \times \sqrt[4]{81}$
$\frac{\sqrt[n]{a}}{\sqrt[n]{b}} = \sqrt[n]{\frac{a}{b}} = \left(\frac{a}{b}\right)^{1/n}$	$\frac{\sqrt{16}}{\sqrt{4}} = \sqrt{4} = 2$
$\sqrt[nx]{a^{mx}} = \sqrt[n]{a^m}$	$\sqrt[8]{a^6} = \sqrt[4]{a^3}$
$\sqrt[n]{a^m} = (\sqrt[n]{a})^m = a^{m/n}$	$\sqrt[4]{a^3} = (\sqrt[4]{a^3}) = a^{3/4}$
$\sqrt{-a} = i\sqrt{a}$	$\sqrt{-25} = i\sqrt{25} = \pm 5i$

Figure 1-30. Formulas associated with roots.

EXPONENTS

Exponents are numbers placed at the upper right hand side of another number or letter. Exponents indicate the raising of some number to a power; that is, they indicate how many times some number or letter is used as a factor. For example, 10^3 means $10 \times 10 \times 10$, and a^3 means $a \times a \times a$. To multiply identical numbers that are expressed in exponential form, add their exponents. See Example 1-32.

Example 1-32: Multiply 2 in the second power by 2 in the fourth power.

Given: $2^2 \times 2^4$

Solution: $a^m \times a^n = a^{m+n}$

$$2^2 \times 2^4 = 2^{2+4} = 2^6 \text{ and}$$
$$2^6 = 2 \times 2 \times 2 \times 2 \times 2 \times 2 = 64$$

Figure 1-31 lists formulas associated with computations involving exponents.

$x \times a^n \pm y \times a^n = (x \pm y)a^n$	$3a^2 + 4a^2 = 7a^2$
$a^m \times a^n = a^{m+n}$	$a^8 \times a^4 = a^{12}$
$\frac{a^m}{a^n} = a^{m-n}$	$\frac{a^8}{a^2} = a^{8-2} = a^6$
$(a^m)^n = (a^n)^m = a^{mn}$	$(a^4)^2 = (a^2)^4 = a^{2 \times 4} = a^8$
$a^{-n} = \frac{1}{a^n}$	$a^{-4} = \frac{1}{a^4}$
$\frac{a^n}{b^n} = \left(\frac{a}{b}\right)^n$	$\frac{a^4}{b^4} = \left(\frac{a}{b}\right)^4$

Figure 1-31. Formulas associated with exponents.

NOTE: Powers and roots are the inverse of each other, and they follow the same algebraic rules. The following is a list of the powers and roots of 10:

$10^0 = 1$

$10^{0.25} = (10)^{1/4} = \sqrt{\sqrt{10}} = \sqrt{3.1623} = 1.778$

$10^{0.33} = (10)^{1/3} = \sqrt[3]{10} = 2.1544$

$10^{0.50} = (10)^{1/2} = \sqrt{10} = 3.1623$

$10^{0.67} = (10)^{2/3} = \sqrt[3]{10^2} = \sqrt[3]{100} = 4.6416$

$10^{0.75} = (10)^{3/4} = \sqrt[4]{(10)^3} = \sqrt{\sqrt{1,000}} = \sqrt{31.623} = 5.623$

$10^{1.0} = 10$

$10^{1.33} = (10)^{4/3} = \sqrt[3]{10^4} = \sqrt[3]{10,000} = 21.544$

$10^{1.50} = (10)^{3/2} = \sqrt{10^3} = \sqrt{1,000} = 31.623$

$10^2 = 100$

$10^{2.33} = (10)^{7/3} = \sqrt[3]{10^7} = \sqrt[3]{10,000,000} = 215.44$

$10^{2.50} = (10)^{5/2} = \sqrt{10^5} = \sqrt{100,000} = 316.23$

$10^3 = 1,000$

$$10^{-1} = \frac{1}{10} = 0.1$$

$$10^{-2} = \frac{1}{(10)^2} = \frac{1}{100} = 0.01$$

$$10^{-3} = \frac{1}{(10)^3} = \frac{1}{1,000} = 0.001$$

TRIGONOMETRY

Trigonometry is the branch of mathematics that involves problem-solving related to triangles. The principles of geometry and algebra are also used in trigonometry.

Problem-solving related to right triangles requires knowledge of the *trigonometric functions.* Problems involving oblique triangles require additional knowledge related to the *Law of Sines* and the *Law of Cosines,* which are discussed later in this chapter.

Trigonometric Functions

Six trigonometric functions exist, each of which is derived from the ratio of two sides of a right triangle and from one of its acute angles. The first three functions, *sine (sin), cosine (cos),* and *tangent (tan)*, are called *primary functions*. These functions can be used to solve most problems related to industrial applications. The other three functions, *cosecant (csc), secant (sec),* and *cotangent (cot)*, are called *secondary functions* (Figure 1-32).

The relative position of these trigonometric functions can be shown in a circle (Figure 1-33). All circles have four *quadrants* (Figure 1-34), each quadrant having an included angle of 90°. The numerical value of each trigonometric function depends on the magnitude of the angle and its sign, and on the position of the angle within the circle.

NOTE: All the trigonometric functions in the first quadrant have a positive sign.

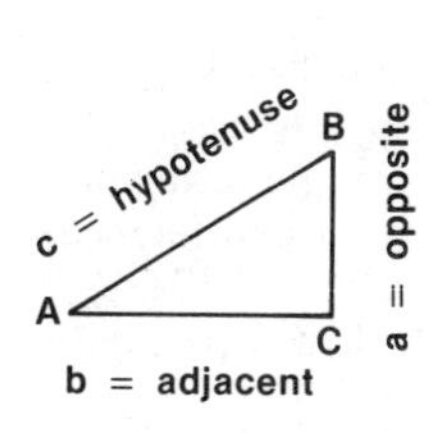

PRIMARY FUNCTIONS	SECONDARY FUNCTIONS
$\sin A = \frac{a}{c} = \frac{\text{opposite side}}{\text{hypotenuse}}$	$\csc A = \frac{c}{a} = \frac{\text{hypotenuse}}{\text{opposite side}}$
$\cos A = \frac{b}{c} = \frac{\text{adjacent side}}{\text{hypotenuse}}$	$\sec A = \frac{c}{b} = \frac{\text{hypotenuse}}{\text{adjacent side}}$
$\tan A = \frac{a}{b} = \frac{\text{opposite side}}{\text{adjacent side}}$	$\cot A = \frac{b}{a} = \frac{\text{adjacent side}}{\text{opposite side}}$

NOTE: THE ANGLES, WHICH ARE OPPOSITE THE SIDES, ARE DESIGNATED A, B, AND C, RESPECTIVELY.

Figure 1-32. Trigonometric function of angle A.

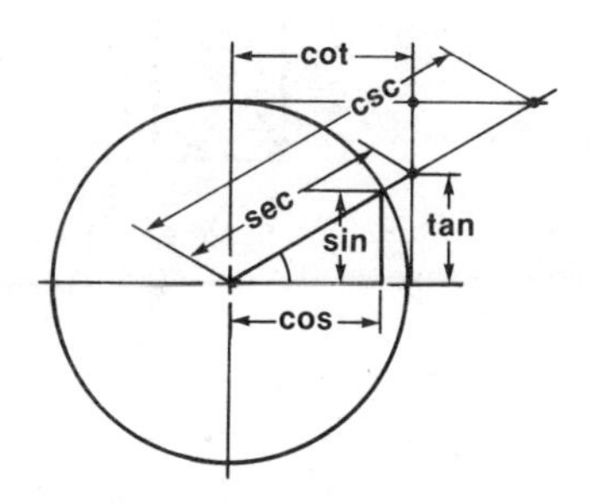

Figure 1-33. Relative position of trigonometric functions for angle A on a circle.

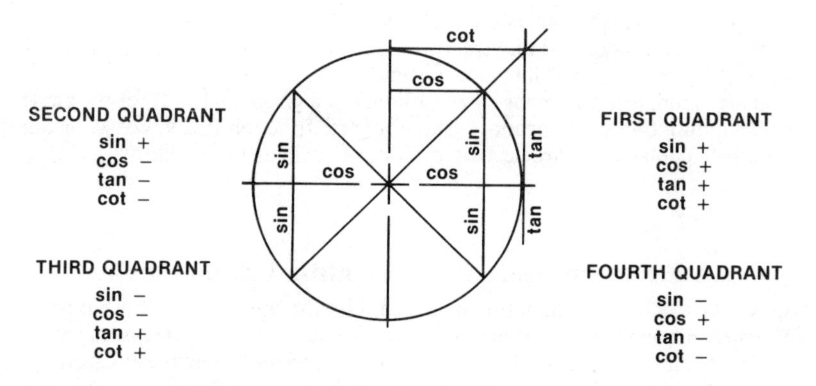

Figure 1-34. Sign of the trigonometric functions in each quadrant.

Radian

In addition to the trigonometric functions and units associated with angles (degrees, minutes, and seconds), another unit, the *radian (rad)*, is frequently used in problems related to trigonometry. A radian is a plane angle subtended by (opposite of) the length of one radius of a given circle on its circumference. There are 6.28 (2π) radians in a circle, where 1 rad = 57.3°, or 57° 17′ 46″. See Figure 1-35 for the relationship between selected angles and their corresponding radians.

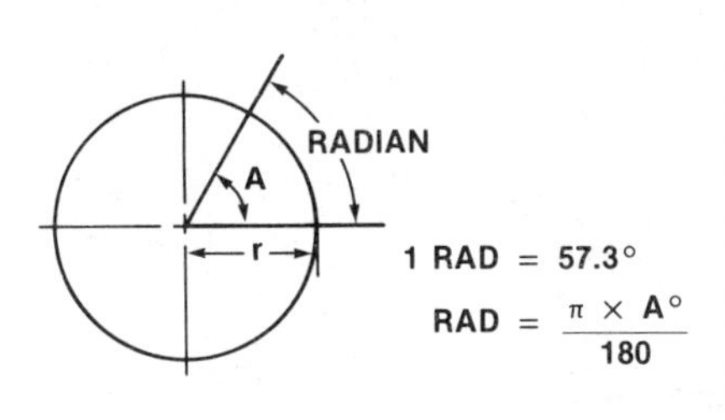

ANGLE	RADIAN	ANGLE	RADIAN
0°	0	90°	$\frac{\pi}{2} = 1.57$
30°	$\frac{\pi}{6} = 0.52$	180°	$\pi = 3.14$
45°	$\frac{\pi}{4} = 0.78$	270°	$\frac{3\pi}{2} = 4.71$
60°	$\frac{\pi}{3} = 1.05$	360°	$2\pi = 6.28$

Figure 1-35. Relationship between angles and radians.

Trigonometric Tables

Trigonometric tables contain the numerical values of the trigonometric functions. Trigonometric functions can be found in tables similar to the one shown in Figure 1-36.

Trigonometric Functions and Calculators

Modern electronic calculators are used for most trigonometric computations. However, problems containing fractions may be computed on a calculator only after they are changed into decimal numbers.

The trigonometric functions corresponding to a given angle (X), or the angle corresponding to a given primary trigonometric function (sin x, cos x, or tan x) can be calculated on an electronic slide rule/scientific calculator. An electronic slide rule/scientific calculator is essential for fast and accurate computations.

Using an Electronic Slide Rule/Scientific Calculator

NOTE: All electronic calculators operate under the same principles, but some differences among the various manufacturer brands exist. For this reason, before using any calculator, it is advisable to read its instruction manual carefully.

The following is a general description of how to use the electronic slide rule/scientific calculator for the computation of the primary trigonometric func-

Degrees	Sin	Cos	Tan	Cot	Sec	Csc	
0° 00′	.0000	1.0000	.0000	——	1.000	——	**90° 00′**
10	029	000	029	343.8	000	343.8	**50**
20	058	000	058	171.9	000	171.9	**40**
30	.0087	1.0000	.0087	114.6	1.000	114.6	**30**
40	116	9999	116	85.94	000	85.95	**20**
50	145	999	145	68.75	000	68.76	**10**
1° 00′	.0175	.9998	.0175	57.29	1.000	57.30	**89° 00′**
10	204	998	204	49.10	000	49.11	**50**
20	233	997	233	42.96	000	42.98	**40**
30	.0262	.9997	.0262	38.19	1.000	38.20	**30**
40	291	996	291	34.37	000	34.38	**20**
50	320	995	320	31.24	001	31.26	**10**
2° 00′	.0349	.9994	.0349	28.64	1.001	28.65	**88° 00′**
10	378	993	378	26.43	001	26.45	**50**
20	407	992	407	24.54	001	24.56	**40**
30	.0436	.9990	.0437	22.90	1.001	22.93	**30**
40	465	989	466	21.47	001	21.49	**20**
50	494	988	495	20.21	001	20.23	**10**
3° 00′	.0523	.9986	.0524	19.08	1.001	19.11	**87° 00′**
10	552	985	553	18.07	002	18.10	**50**
20	581	983	582	17.17	002	17.20	**40**
30	.0610	.9981	.0612	16.35	1.002	16.38	**30**
40	640	980	641	15.60	002	15.64	**20**
50	669	978	670	14.92	002	14.96	**10**
4° 00′	.0698	.9976	.0699	14.30	1.002	14.34	**86° 00′**
10	727	974	729	13.73	003	13.76	**50**
20	756	971	758	13.20	003	13.23	**40**
30	.0785	.9969	.0787	12.71	1.003	12.75	**30**
40	814	967	816	12.25	003	12.29	**20**
50	843	964	846	11.83	004	11.87	**10**
5° 00′	.0872	.9962	.0875	11.43	1.004	11.47	**85° 00′**
10	901	959	904	11.06	004	11.10	**50**
20	929	957	934	10.71	004	10.76	**40**
30	.0958	.9954	.0963	10.39	1.005	10.43	**30**
40	987	951	992	10.08	005	10.13	**20**
50	.1016	948	.1022	9.788	005	9.839	**10**
6° 00′	.1045	.9945	.1051	9.514	1.006	9.567	**84° 00′**
10	074	942	080	9.255	006	9.309	**50**
20	103	939	110	9.010	006	9.065	**40**
30	.1132	.9936	.1139	8.777	1.006	8.834	**30**
40	161	932	169	8.556	007	8.614	**20**
50	190	929	198	8.345	007	8.405	**10**
7° 00′	.1219	.9925	.1228	8.144	1.008	8.206	**83° 00′**
10	248	922	257	7.953	008	8.016	**50**
20	276	918	287	7.770	008	7.834	**40**
30	.1305	.9914	.1317	7.596	1.009	7.661	**30**
40	334	911	346	7.429	009	7.496	**20**
50	363	907	376	7.269	009	7.337	**10**
8° 00′	.1392	.9903	.1405	7.115	1.010	7.185	**82° 00′**
10	421	899	435	6.968	010	7.040	**50**
20	449	894	465	6.827	011	6.900	**40**
30	.1478	.9890	.1495	6.691	1.011	6.765	**30**
40	507	886	524	6.561	012	6.636	**20**
50	536	881	554	6.435	012	6.512	**10**
9° 00′	.1564	.9877	.1584	6.314	1.012	6.392	**81° 00′**
	Cos	**Sin**	**Cot**	**Tan**	**Csc**	**Sec**	**Degrees**

Figure 1-36. Trigonometric table for the numerical values of the trigonometric functions of angles 0°–9° and 81°–90°.

tions of an angle, and the angle that corresponds to a certain trigonometric function.

To find the primary trigonometric functions of a given angle:

1. Enter the numerical value of a given angle in degrees and decimal of a degree.
2. Depress the desired function key (sin, cos, or tan) and the calculator's display will show the value.

To find the angle of a given primary trigonometric function:

1. Enter the numerical value of the given trigonometric function.
2. Depress the INV (inverse) key.
3. Depress the key of the given trigonometric function and the calculator's display will show the value of the angle in degrees and decimal of a degree.

Solving Problems Related to Right Triangles

Any problem related to right triangles can be solved by using one of the formulas in Figure 1-37.

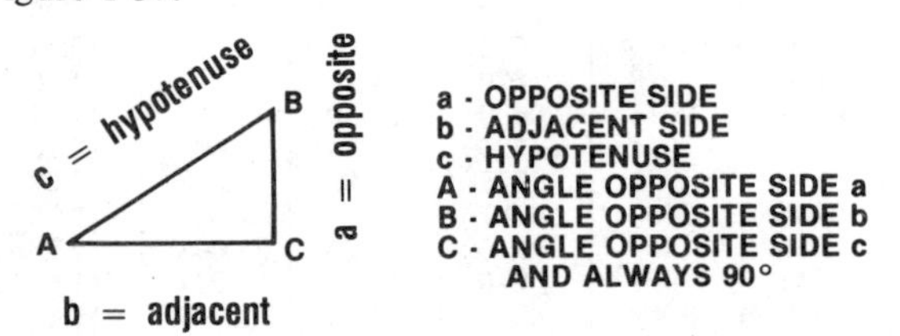

KNOWN	FORMULAS FOR THE UNKNOWN		
Sides a and c	$b = \sqrt{c^2 - a^2}$	$\sin A = \frac{a}{c}$	$B = 90° - A$
Sides b and c	$a = \sqrt{c^2 - b^2}$	$\sin B = \frac{b}{c}$	$A = 90° - B$
Sides a and b	$c = \sqrt{a^2 + b^2}$	$\tan A = \frac{a}{b}$	$B = 90° - A$
Side c and angle A	$a = c \times \sin A$	$b = c \times \cos A$	$B = 90° - A$
Side c and angle B	$a = c \times \cos B$	$b = c \times \sin B$	$A = 90° - B$
Side a and angle A	$c = \frac{a}{\sin A}$	$b = a \times \cot A$	$B = 90° - A$
Side a and angle B	$c = \frac{a}{\cos B}$	$b = a \times \tan B$	$A = 90° - B$
Side b and angle A	$c = \frac{b}{\cos A}$	$a = b \times \tan A$	$B = 90° - A$
Side b and angle B	$c = \frac{b}{\sin B}$	$a = b \times \cot B$	$A = 90° - B$

Figure 1-37. Formulas for solving right-triangle problems.

In order to solve a problem related to a right triangle, either two of its sides, or one side and one acute angle, must be known. See Examples 1-33 and 1-34.

Example 1-33: Find the acute angles A and B, and the adjacent side b of a right triangle that has a hypotenuse of 4″ and opposite side of 2″.

Solution:

Angle A: $\sin A = \frac{a}{c} = \frac{2}{4} = \frac{1}{2} = 0.500$

$A = 30°$

Angle B: $90° - A = 90° - 30° = 60°$

Side b: $\cos A = \frac{b}{c}$ and $b = \cos A \times c$

$\cos A = 0.866$ and
$b = 0.866 \times 4 = 3.464''$

NOTE: The trigonometric function that corresponds to a certain angle, or to the angle that corresponds to a certain function, can be found by using an electronic slide rule/scientific calculator, or a trigonometric table. In this case, the angle of sine .500 is 30°, and the cosine of 30° is .866.

Example 1-34: Find the hypotenuse c of a right triangle that has an opposite side of 6.5″ and an angle B of 72°.

Solution: Angle $A = 90° - B = 90° - 72° = 18°$

$\sin 18° = .309$

$\sin A = \frac{a}{c}$ and $c = \frac{a}{\sin A} = \frac{6.5}{.309} = 21.036''$

Solving Problems Related to Oblique Triangles

Any problem related to oblique triangles (Figure 1-38) can be solved by using the formulas derived from the *Law of Cosines* and the *Law of Sines*.

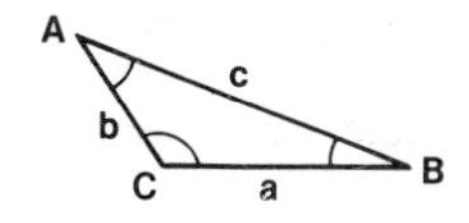

Figure 1-38. Oblique triangle.

Law of Cosines. The Law of Cosines states that in any triangle, the square of the unknown side equals the sum of the squares of the given sides minus two times the product of the given sides, times the cosine of the angle included between them. That is,

$$a^2 = b^2 + c^2 - 2bc(\cos A) \text{ or } \cos A = \frac{a^2 + b^2 + c^2}{2bc}$$

$$b^2 = c^2 + a^2 - 2ca(\cos B) \text{ or } \cos B = \frac{a^2 + b^2 + c^2}{2ac}$$

$$c^2 = a^2 + b^2 - 2ab(\cos C) \text{ or } \cos C = \frac{a^2 + b^2 + c^2}{2bc}$$

The Law of Cosines is used to solve any problem if two sides and the included angle between them are known. See Example 1-35.

Law of Sines. The Law of Sines states that in any triangle, the sines of the angles are proportional to the sides opposite them. That is,

$$\frac{\sin A}{a} = \frac{\sin B}{b} = \frac{\sin C}{c}$$

The Law of Sines is used to solve any problem if two sides and an angle are known, or if two angles and one side are known. See Example 1-36.

Example 1-35: Find the length of side a of an oblique triangle using the formulas derived from the law of cosines.

Given: Side $b = 5''$, Side $c = 10''$, Angle $A = 50°$, $\cos 50° = 0.6428$

Solution: $a^2 = b^2 + c^2 - 2bc \times \cos A$

$$= 5^2 + 10^2 - 2 \times 5 \times 10 \times 0.6428$$

$$a^2 = 25 + 100 - 100 \times 0.6428 = 125 - 64.28 = 60.72$$

$$a = \sqrt{60.72} = 7.8''$$

Example 1-36: Find the angle B of an oblique triangle using the formulas derived from the law of sines.

Given: side $a = 30''$, side $b = 20''$, angle $A = 55°$, $\sin 55° = 0.8192$

Solution: $\frac{\sin B}{b} = \frac{\sin A}{a}$ and $\sin B = \frac{b \times \sin A}{a}$

$$\sin B = \frac{20 \times 0.8192}{30} = \frac{16.384}{30} = 0.5461$$

$$\sin B = 0.5461 \text{ and } B = 33.1°$$

Chapter 2

UNITS OF MEASUREMENT

Measurement is the process used to find amounts. A measurement is found by comparing the measuring quantity with a reference quantity of the same kind (a standard unit).

Two systems of measurement are commonly used: (1) the International System of Units (SI metric), which is used in all countries, and (2) the English/American customary system which is used mainly in the United States. Other English-speaking countries have replaced the English system of units with the SI metric system of units.

The SI metric system of units is a modernized version of the "old" metric system of weights and measures. The new metric system was adopted in 1960 as the International System of Units.

INTERNATIONAL SYSTEM OF UNITS (SI METRIC)

The SI metric system of measurement consists of: (a) 7 base units, (b) 2 supplementary units, (c) decimal numerical prefixes for the formation of new units representing multiples or sub-multiples of the base units, and (d) derived units.

Base Units

NAME OF UNIT	SI SYMBOL	QUANTITY
meter	m	length
kilogram	kg	mass
second	s	time
kelvin	K	thermodynamic temperature
ampere	A	electric current
candela	cd	luminous intensity
mole	mol	amount of substance

Supplementary Units

NAME OF UNIT	SI SYMBOL	QUANTITY
radian	rad	plane angle
steradian	sr	solid angle

Prefixes of Units

MULTIPLES AND SUBMULTIPLES	PREFIXES	SYMBOLS	MEANING
1,000,000,000,000 = 10^{12}	tera	T	trillion
1,000,000,000 = 10^{9}	giga	G	billion
1,000,000 = 10^{6}	mega	M	million
1,000 = 10^{3}	kilo	k	thousand
100 = 10^{2}	hecto	h	hundred
10 = 10^{1}	deka	d	ten
Unit 1 = 10^{0}			
0.1 = 10^{-1}	deci	d	tenth
0.01 = 10^{-2}	centi	c	hundredth
0.001 = 10^{-3}	milli	m	thousandth
0.000001 = 10^{-6}	micro	μ	millionth
0.000000001 = 10^{-9}	nano	n	billionth
0.000000000001 = 10^{-12}	pico	p	trillionth

NOTE: Each of these prefixes may be used to describe a new unit that is a power of ten of the original unit. For example, centi (c), the prefix associated with 10^{-2}, may form the centimeter (cm), which is a unit equal to one hundredth of a meter (0.01 m); or the prefix kilo (10^3) may form the kilometer (km) which is a unit equal to one thousand meters (1,000 m).

Derived Units with Special Names

UNIT	SYMBOL	QUANTITY	RELATION TO OTHER UNITS
coulomb	C	electrical charge, electrical potential	C = As
farad	F	capacitance	F = C/V
henry	H	inductance	H = Wb/m^2
hertz	Hz	frequency	Hz = s^{-1}
joule	J	energy, work, quantity of heat	J = N·m
newton	N	force	N = $kg\ m/s^2$
ohm	Ω	electrical resistance	Ω = V/A
pascal	Pa	pressure	Pa = N/m^2
siemens	S	conductance	S = A/V
tesla	T	magnetic flux density	T = Wb/m^2
volt	V	potential difference electromotive force	V = W/A
watt	W	power, radiant flux quantity of electricity	W = J/S
weber	Wb	magnetic flux	Wb = V·s

NOTE: Lower case symbol s represents time (second).

Derived Units Having Algebraic Relationships with Base Units and Units with Special Names

UNIT	SYMBOL	QUANTITY
meter-per-second squared	m/s^2	acceleration
radian-per-second squared	rad/s^2	angular acceleration
radian per second	rad/s	angular velocity
mole per cubic meter	mol/m^3	concentration
ampere per square meter	A/m^2	current density
kilogram per cubic meter	kg/m^3	density
coulomb per meter	C/m	electrical charge density
coulomb per square meter	C/m^2	electrical flux density
joule per cubic meter	J/m^3	energy density
watt per square meter	W/m^2	heat flux density
joule per kelvin	J/K	heat capacity, entropy
square meter per second	m^2/s	kinematic viscosity
candela per square meter	cd/m^2	luminance
ampere per meter	A/m	magnetic field strength
joule per mole	J/mol	molar energy
joule per mole kelvin	J/mol °K	molar entropy molar heat capacity
newton meter	N·m	movement of force
henry per meter	H/m	permeability (magnetic)
farad per meter	F/m	permittivity
watt per square meter steradian	$W/m^2 \cdot sr$	radiance
watt per steradian	W/sr	radiant intensity
joule per kilogram	J/kg	specific energy
joule per kilogram kelvin	J/(kg·K)	specific heat capacity specific entropy
newton per square meter	N/m^2	stress
newton per meter	N/m	surfaçe tension
watt per meter kelvin	W/(m·K)	thermal conductivity
meter per second	m/s	velocity
newton second per square meter	$N \cdot s/m^2$	viscosity

Rules Concerning the Use of SI Units

1. Every unit is represented by its accepted symbol without a period and only in the singular form.
2. The full names of units are always written in lower case letters except at the beginning of a sentence.
3. The symbols for SI units are written in lower case letters except for Kelvin (K), ampere (A), the derived units with special names, and the prefixes for: 10^6 (M), 10^9 (G), and 10^{12} (T).
4. Prefixes are used when expressing numbers less than 10^{-3} (.001) and greater than 10^3 (1,000).
5. Space is provided between a number and the sumbol and to separate long numbers in three digit intervals.

DEFINITION OF SI METRIC UNITS AND THEIR RELATIONSHIP TO ENGLISH UNITS

The SI base unit of length is the meter (m). The unit for measuring area is the square meter (m^2); the unit for measuring volume is the cubic meter (m^3); and the unit for measuring liquid volume is the cubic decimeter (dm^3).

Units of Linear Measurement

SI METRIC UNITS OF LENGTH	ENGLISH UNITS OF LENGTH
meter (m)	yard (yd)
1 decimeter (dm) = 0.1 m	1 foot = $\frac{1}{3}$ yd = 12″
1 centimeter (cm) = 0.01 m	1 inch = $\frac{1}{36}$ yd = $\frac{1'}{12}$
1 millimeter (mm) = 0.001 m	
1 m = 10 dm = 100 cm = 1,000 mm	1 yd = 3′ = 36″
kilometer (km)	rod = 5.5 yd
1 kilometer = 1,000 m	mile = 1,760 yd

Relationship Between Metric and English Units of Length

LENGTH	meter (m)	millimeter (mm)	foot (ft)	inch
1 meter	1	1,000	3.281	39.37
1 millimeter	0.001	1	3.281×10^{-3}	0.03937
1 foot	0.3048	304.8	1	12
1 inch	0.0254	25.4	0.0833	1

Conversion of Units of Length. The conversion of units is based upon any two units of the same quantity, regardless of the system of measurement, having a definite relationship between themselves. This relationship may be expressed as a product of one unit times its equivalent (product method), or as a fraction of the smaller unit over its equivalent (ratio method). (The fraction is the ratio between the two units.) For example, the relationship between the inch and the millimeter is 1″ = 25.4 mm or 1 mm = 0.03937″. See Example 2-1.

Example 2-1: Convert 200 mm into inches by using both the product and the ratio method.

Solution: Product Method | Ratio Method

$$200 \times 0.03937 = 7.874''$$

$$200 \times \frac{1}{25.4} = \frac{200}{25.4} = 7.874''$$

In problems where different units of the same quantity are used, it is advisable to write the smaller unit as a ratio instead of converting it. See Example 2-2.

Example 2-2: The length of a rectangular plane is 435 mm and the width is 18″. Find the area of this plane in square inches.

Solution: $A = L \times W = \frac{435}{25.4} \times 18 = \frac{7{,}830}{25.4} = 308.267 \text{ in}^2$

Units of Area or Square Measurement

SI METRIC UNITS OF AREA	ENGLISH UNITS OF AREA
square meter (m^2)	square yard (yd^2) = 9 ft^2
square decimeter (dm^2) = 0.01 m^2	square foot (ft^2) = 144 in^2
square centimeter (cm^2) = 0.0001 m^2	square inch (in^2)
square millimeter (mm^2) = 0.000001 m^2	square rod = 30.25 yd^2
square kilometer (km^2) = 1,000,000 m^2	acre = 4,840 yd^2
hectare = 1,000 m^2	square mile = 640 acres

Relationship Between Metric and English Units of Area

AREA	m^2	cm^2	ft^2	in^2
1 square meter	1	10^4	10.764	1,549.9
1 square centimeter	10^{-4}	1	0.00108	0.155
1 square foot	0.0929	929	1	144
1 square inch	6.452×10^{-4}	6.452	0.00694	1

Conversions of Units of Area

Example 2-3: Convert 24 square feet (24 ft^2) into square meters (m^2).
Solution: $24 \times 0.0929 = 2.23\ m^2$ or

$$24 \times \frac{1}{10.764} = \frac{24}{10.764} = 2.23\ m^2$$

Example 2-4: Convert 85 square centimeters (85 cm^2) into square inches (in^2).
Solution: $85 \times 0.155 = 13.175\ in^2$ or

$$85 \times \frac{1}{6.452} = \frac{85}{6.452} = 13.175\ in^2$$

Units of Solid Volume

SI METRIC UNITS OF VOLUME	ENGLISH UNITS OF SOLID VOLUME
cubic meter (m^3)	cubic yard (yd^3) = 27 ft^3
cubic decimeter (dm^3) = 0.001 m^3	cubic foot (ft^2) = 1,728 in^3
cubic centimeter (cm^3) = 0.000001 m^3	cubic inch (in^3)
cubic millimeter (mm^3) = 0.000000001 m^3	

Relationship Between Metric and English Units of Solid Volume

VOLUME	m^3	dm^3	cm^3	in^3	ft^3
1 cubic meter	1	1,000	10^6	61,023	35.31
1 cubic decimeter	10^{-4}	1	10^4	61.023	0.0307
1 cubic centimeter	10^{-7}	0.001	1	0.061023	3.531×10^{-5}
1 cubic foot	0.028317	28.317	28,320	1,728	1
1 cubic inch	1.639×10^{-5}	0.01639	16.39	1	5.787×10^{-4}

Units of Liquid Volume

SI METRIC UNITS OF LIQUID VOLUME	ENGLISH UNITS OF LIQUID VOLUME
liter = 1 dm^3 milliliter = 1 cm^3 = 0.001 liter kiloliter = 1,000 dm^3 or 1,000 liters	cubic foot (ft^3) cubic inch (in^3) gallon: 1 gallon = 4 quarts quart: 1 quart = 2 pints pint: 1 pint = 16 ounces

Relationship Between Metric and English Units of Liquid Volume

1 liter = 0.0353154 ft^3	1 ft^3 = 28.31622 liters
1 liter = 0.264178 gallon	1 in^3 = 16.387064 milliliters
1 cm^3 = 0.0610234 in^3	1 gallon = 3.785 liters
	1 quart = 0.946 liter
	1 pint = 0.473 liter
	1 ounce = 29.5625 milliliters

Conversion of Units of Solid and Liquid Volume

Example 2-5: Convert 50 liters into gallons.
Solution: $50 \times 0.264 = 13.2$ gal or

$$50 \times \frac{1}{3.785} = 13.21 \text{ gal}$$

Example 2-6: Convert 825 cubic milliliters (825 cm^3) into cubic inches (in^3).
Solution: $825 \times 0.061 = 50.325$ in^3 or

$$825 \times \frac{1}{16.387} = 50.344 \text{ in}^3$$

NOTE: By rounding off the equivalent of a particular unit as in Examples 2-5 and 2-6, the result of the conversion may be slightly larger or slightly smaller depending on how it is rounded off. For this reason, one should use his or her best judgement before rounding off any equivalent unit.

MASS

The SI metric base unit of mass is the kilogram (kg). The standard (prototype) of this unit is a cylinder made of platinum-iridium alloy and is kept in Paris, France at the International Bureau of Weights and Measures.

Units of Mass

SI METRIC UNITS OF MASS	ENGLISH UNITS OF MASS
1 kilogram (kg) = 1,000 g	1 pound (lb) = 16 oz
1 gram (g) = 0.001 kg	1 ounce (oz) = $\frac{1}{16}$ lb
1 metric ton (t) = 1,000 kg	1 slug = 32.1 lb 1 short ton = 2,000 lb 1 long ton = 2,240 lb

Relationship Between Metric and English Units of Mass

MASS	g	kg	lb	sl
1 gram	1	0.001	0.002205	6.852×10^{-5}
1 kilogram	1,000	1	2.204	0.06852
1 pound	453.6	0.4536	1	0.03108
1 slug	14,590	14.590	32.17	1

Conversions of Units of Mass

Example 2-7: Convert 25 kilograms (25 kg) into pounds (lb).
Solution: $25 \times 2.204 = 55$ lb or

$$25 \times \frac{1}{0.4536} = \frac{25}{0.4536} = 55.1 \text{ lb}$$

Example 2-8: Convert 48 ounces (48 oz) into grams (g).
NOTE: 1 ounce = 28.35 grams and the reciprocal of 1 ounce is 0.0352.

Solution: $48 \times \frac{453.6}{16} = 48 \times 28.35 = 1{,}360.8$ g or

$$48 \times \frac{1}{0.0352} = \frac{48}{0.0352} = 1{,}363.6 \text{ g}$$

TIME

The SI base unit of time is the second (s). A second is a scientifically measured duration of time based upon cycles of radiation measured with a spectrometer.

Units of Time and Their Relationship

seconds (s)	1 min = 60 s
minute (min)	1 h = 60 min
hour (h)	1 d = 24 h
day (d)	1 y = 365.26 d
year (y)	

THERMODYNAMIC TEMPERATURE

The SI base unit of temperature is the kelvin (K). The degree kelvin (°K) or degree of absolute temperature is used mainly for scientific measurements. For practical applications, the unit of temperature used most often is the degree Celsius (°C). The degree Celsius is identical to the degree kelvin, but the scale that measures in degrees Celsius is more convenient to use.

Units of Temperature

SI METRIC UNITS OF TEMPERATURE	ENGLISH UNITS OF TEMPERATURE
degrees Kelvin (°K) degrees Celsius (°C)	degrees Fahrenheit (°F)

Temperature Scales and Their Relationship

SCALE	ABSOLUTE ZERO	ICE POINT	BOILING WATER
Kelvin	0°K	273.16°K	373.15°K
Celsius	−273.16°C	0°C	100°C
Fahrenheit	−459.69°F	32°F	212°F

Conversion of Units of Temperature. The following formulas are used to convert units of temperature from one scale to the other:

1. $t°K = t°C + 273.16$ or $t°K = t°F + 459.69$

2. $t°C = \frac{5}{9}(t°F - 32)$

3. $t°F = \frac{9}{5}(t°C + 32)$

NOTE: For examples of temperature conversion, see Chapter 4, Units of Temperature, Examples 4-8 and 4-9.

ELECTRICAL CURRENT

The SI base unit of electrical current is the ampere (A). An ampere is equal to one coulomb (C) per second past a given point in a circuit.

Units of Electrical Current and Their Relationship

ampere (A)	$V = \frac{W}{A}$
volt (V)	$\Omega = \frac{V}{A}$
ohm (Ω)	

NOTE: The letter W indicates the derived unit watt.

LUMINOUS INTENSITY

The SI base unit of luminous intensity is the candela (cd), which is derived from exact scientific measurements.

Units of Luminous Intensity

SI METRIC UNITS OF LUMINOUS INTENSITY	ENGLISH UNITS OF LUMINOUS INTENSITY
candela (cd) - luminous intensity lumen (lm) - luminous flux lux (lx) - illumination cd = 4 × π × lm × lx = lm/m²	foot candle (fc) foot lambert (fl) fc = lx(1.0764) fl = cd/m²(3.4262)

NOTE: Common light bulbs emit approximately 170 lumens per 10W of electrical measurement.

AMOUNT OF SUBSTANCE

The SI base unit for amount of substance is the mole (mol). The mole is a quantity of chemical substance expressed in grams. For example, a mole of hydrogen (H_2) gas is 2 grams while a mole of oxygen (O_2) gas is 32 grams.

RADIAN

The radian (rad) is a plane angle with its vertex at the center of a circle that is subtended by (opposite of) an arc equal in length to the radius (Figure 2-1). The included angle in a radian corresponds to 57° 17′ 45″, or 57.3°. (See chapter 1 under TRIGONOMETRY for additional radian information.)

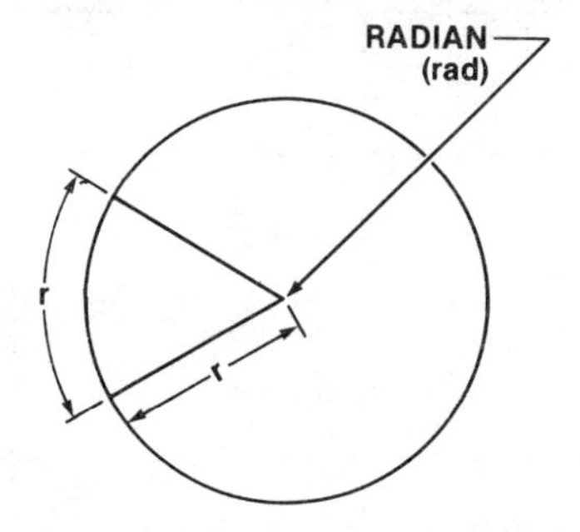

Figure 2-1. Radian (plane angle).

STERADIAN

The steradian (sr) is a solid angle (Figure 2-2) with its vertex at the center of a sphere that is subtended by an area of the spherical surface equal to that of a square with sides equal in length to the radius.

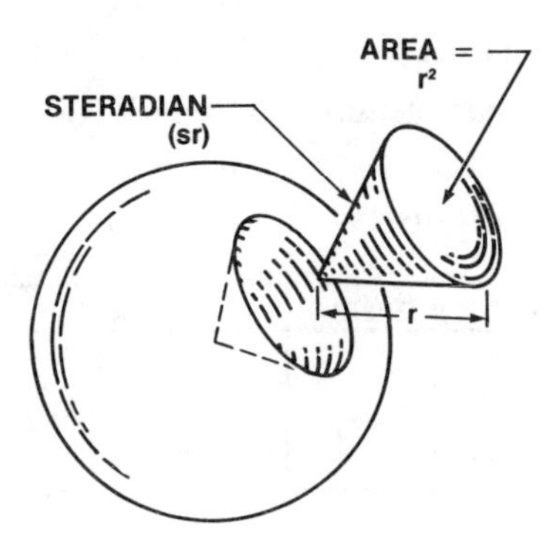

Figure 2-2. Steradian (solid angle).

Common Engineering Units and Their Relationship

QUANTITY	SI METRIC UNITS/SYMBOLS	CUSTOMARY UNITS	RELATIONSHIP OF UNITS
Acceleration	meters per second squared (m/s^2)	feet per second (ft/s^2)	$m/s^2 = ft/s^2 \times 3.281$
Area	square meter (m^2) square millimeter (mm^2)	square foot (ft^2) square inch (in^2)	$m^2 = ft^2 \times 10.764$ $mm^2 = in^2 \times 0.00155$
Density	kilograms per cubic meter (kg/m^3) grams per cubic centimeter (g/cm^3)	pounds per cubic foot (lb/ft^3) pounds per cubic inch (lb/in^3)	$kg/m^3 = lb/ft^3 \times 16.02$ $g/cm^3 = lb/in^3 \times 0.036$
Work	Joule (J)	foot pound force (ft lbf or ft lb)	J = ft lbf × 1.356
Heat	Joule (J)	British thermal unit (Btu) Calorie (Cal)	J = Btu × 1.055 J = cal × 4.187
Energy	kilowatt (kW)	Horsepower (HP)	kW = HP × 0.7457
Force	Newton (N) Newton (N)	Pound-force (lbf, lb·f, or lb) kilogram-force (kgf, kg·f, or kp)	N = lbf × 4.448 $N = \frac{kgf}{9.807}$
Length	meter (m) millimeter (mm)	foot (ft) inch (in)	m = ft × 3.281 $mm = \frac{in}{25.4}$
Mass	kilogram (kg) gram (g)	pound (lb) ounce (oz)	kg = lb × 2.2 $g = \frac{oz}{28.35}$
Stress	Pascal = Newton per second (Pa = N/s)	pounds per square inch (lb/in^2 or psi)	$Pa = lb/in^2 \times 6{,}895$
Temperature	degree Celsius (°C)	degree Fahrenheit (°F)	$°C = \frac{°F - 32}{1.8}$
Torque	Newton meter (N·m)	foot-pound (ft lb) inch-pound (in lb)	N·m = ft lbf × 1.356 N·m = in lbf × 0.113
Volume	cubic meter (m^3) cubic centimeter (cm^3)	cubic foot (ft^3) cubic inch (in^3)	$m^3 = ft^3 \times 35.314$ $cm^3 = \frac{in^3}{16.387}$

Chapter 3

MACHINE TOOL TECHNOLOGY

The purpose of machining operations is to produce parts of specified shapes and dimensions. Machine shop work requires familiarity with basic measuring tools, and basic principles of machining metals and other engineering materials.

MEASURING TOOLS

Measuring tools commonly used in machine shop work include the *steel rule, vernier caliper, micrometer,* and *vernier protractor*. Choosing the proper tool for a certain measurement depends upon the size, shape, and desired accuracy of the part being machined.

Measuring tools are graduated in English units, metric units, or both. English measuring tools use the inch as a unit of measurement while metric measuring tools use the millimeter (mm).

Steel Rule

The steel rule (Figure 3-1) is the most commonly used measuring tool for semi-

Figure 3-1. Steel rule graduations. (The L. S. Starrett Company)

units. English steel rules are graduated in inches and fractions of an inch, such as eighths, sixteenths, thirty-seconds, and sixty-fourths. Some are graduated in tenths and fiftieths of an inch. Metric steel rules are graduated in millimeters and in halves of a millimeter. See Figures 3-2 and 3-3 for English and metric equivalents.

DECIMALS TO MILLIMETERS

Decimal	mm
0.001	0.0254
0.002	0.0508
0.003	0.0762
0.004	0.1016
0.005	0.1270
0.006	0.1524
0.007	0.1778
0.008	0.2032
0.009	0.2286
0.010	0.2540
0.020	0.5080
0.030	0.7620
0.040	1.0160
0.050	1.2700
0.060	1.5240
0.070	1.7780
0.080	2.0320
0.090	2.2860
0.100	2.5400
0.110	2.7940
0.120	3.0480
0.130	3.3020
0.140	3.5560
0.150	3.8100
0.160	4.0640
0.170	4.3180
0.180	4.5720
0.190	4.8260
0.200	5.0800
0.210	5.3340
0.220	5.5880
0.230	5.8420
0.240	6.0960
0.250	6.3500
0.260	6.6040
0.270	6.8580
0.280	7.1120
0.290	7.3660
0.300	7.6200
0.310	7.8740
0.320	8.1280
0.330	8.3820
0.340	8.6360
0.350	8.8900
0.360	9.1440
0.370	9.3980
0.380	9.6520
0.390	9.9060
0.400	10.1600
0.410	10.4140
0.420	10.6680
0.430	10.9220
0.440	11.1760
0.450	11.4300
0.460	11.6840
0.470	11.9380
0.480	12.1920
0.490	12.4460

Decimal	mm
0.500	12.7000
0.510	12.9540
0.520	13.2080
0.530	13.4620
0.540	13.7160
0.550	13.9700
0.560	14.2240
0.570	14.4780
0.580	14.7320
0.590	14.9860
0.600	15.2400
0.610	15.4940
0.620	15.7480
0.630	16.0020
0.640	16.2560
0.650	16.5100
0.660	16.7640
0.670	17.0180
0.680	17.2720
0.690	17.5260
0.700	17.7800
0.710	18.0340
0.720	18.2880
0.730	18.5420
0.740	18.7960
0.750	19.0500
0.760	19.3040
0.770	19.5580
0.780	19.8120
0.790	20.0660
0.800	20.3200
0.810	20.5740
0.820	20.8280
0.830	21.0820
0.840	21.3360
0.850	21.5900
0.860	21.8440
0.870	22.0980
0.880	22.3520
0.890	22.6060
0.900	22.8600
0.910	23.1140
0.920	23.3680
0.930	23.6220
0.940	23.8760
0.950	24.1300
0.960	24.3840
0.970	24.6380
0.980	24.8920
0.990	25.1460
1.000	25.4000

FRACTIONS TO DECIMALS TO MILLIMETERS

Fraction	Decimal	mm	Fraction	Decimal	mm
1/64	0.0156	0.3969	33/64	0.5156	13.0969
1/32	0.0312	0.7938	17/32	0.5312	13.4938
3/64	0.0469	1.1906	35/64	0.5469	13.8906
1/16	0.0625	1.5875	9/16	0.5625	14.2875
5/64	0.0781	1.9844	37/64	0.5781	14.6844
3/32	0.0938	2.3812	19/32	0.5938	15.0812
7/64	0.1094	2.7781	39/64	0.6094	15.4781
1/8	0.1250	3.1750	5/8	0.6250	15.8750
9/64	0.1406	3.5719	41/64	0.6406	16.2719
5/32	0.1562	3.9688	21/32	0.6562	16.6688
11/64	0.1719	4.3656	43/64	0.6719	17.0656
3/16	0.1875	4.7625	11/16	0.6875	17.4625
13/64	0.2031	5.1594	45/64	0.7031	17.8594
7/32	0.2188	5.5562	23/32	0.7188	18.2562
15/64	0.2344	5.9531	47/64	0.7344	18.6531
1/4	0.2500	6.3500	3/4	0.7500	19.0500
17/64	0.2656	6.7469	49/64	0.7656	19.4469
9/32	0.2812	7.1438	25/32	0.7812	19.8438
19/64	0.2969	7.5406	51/64	0.7969	20.2406
5/16	0.3125	7.9375	13/16	0.8125	20.6375
21/64	0.3281	8.3344	53/64	0.8281	21.0344
11/32	0.3438	8.7312	27/32	0.8438	21.4312
23/64	0.3594	9.1281	55/64	0.8594	21.8281
3/8	0.3750	9.5250	7/8	0.8750	22.2250
25/64	0.3906	9.9219	57/64	0.8906	22.6219
13/32	0.4062	10.3188	29/32	0.9062	23.0188
27/64	0.4219	10.7156	59/64	0.9219	23.4156
7/16	0.4375	11.1125	15/16	0.9375	23.8125
29/64	0.4531	11.5094	61/64	0.9531	24.2094
15/32	0.4688	11.9062	31/32	0.9688	24.6062
31/64	0.4844	12.3031	63/64	0.9844	25.0031
1/2	0.5000	12.7000	1	1.0000	25.4000

Figure 3-2. Equivalents of English and Metric units. (The L. S. Starrett Company)

MILLIMETERS TO DECIMALS

mm	Decimal	mm	Decimal	mm	Decimal	mm	Decimal	mm	Decimal
0.01	.00039	0.41	.01614	0.81	.03189	21	.82677	61	2.40157
0.02	.00079	0.42	.01654	0.82	.03228	22	.86614	62	2.44094
0.03	.00118	0.43	.01693	0.83	.03268	23	.90551	63	2.48031
0.04	.00157	0.44	.01732	0.84	.03307	24	.94488	64	2.51969
0.05	.00197	0.45	.01772	0.85	.03346	25	.98425	65	2.55906
0.06	.00236	0.46	.01811	0.86	.03386	26	1.02362	66	2.59843
0.07	.00276	0.47	.01850	0.87	.03425	27	1.06299	67	2.63780
0.08	.00315	0.48	.01890	0.88	.03465	28	1.10236	68	2.67717
0.09	.00354	0.49	.01929	0.89	.03504	29	1.14173	69	2.71654
0.10	.00394	0.50	.01969	0.90	.03543	30	1.18110	70	2.75591
0.11	.00433	0.51	.02008	0.91	.03583	31	1.22047	71	2.79528
0.12	.00472	0.52	.02047	0.92	.03622	32	1.25984	72	2.83465
0.13	.00512	0.53	.02087	0.93	.03661	33	1.29921	73	2.87402
0.14	.00551	0.54	.02126	0.94	.03701	34	1.33858	74	2.91339
0.15	.00591	0.55	.02165	0.95	.03740	35	1.37795	75	2.95276
0.16	.00630	0.56	.02205	0.96	.03780	36	1.41732	76	2.99213
0.17	.00669	0.57	.02244	0.97	.03819	37	1.45669	77	3.03150
0.18	.00709	0.58	.02283	0.98	.03858	38	1.49606	78	3.07087
0.19	.00748	0.59	.02323	0.99	.03898	39	1.53543	79	3.11024
0.20	.00787	0.60	.02362	1.00	.03937	40	1.57480	80	3.14961
0.21	.00827	0.61	.02402	1	.03937	41	1.61417	81	3.18898
0.22	.00866	0.62	.02441	2	.07874	42	1.65354	82	3.22835
0.23	.00906	0.63	.02480	3	.11811	43	1.69291	83	3.26772
0.24	.00945	0.64	.02520	4	.15748	44	1.73228	84	3.30709
0.25	.00984	0.65	.02559	5	.19685	45	1.77165	85	3.34646
0.26	.01024	0.66	.02598	6	.23622	46	1.81102	86	3.38583
0.27	.01063	0.67	.02638	7	.27559	47	1.85039	87	3.42520
0.28	.01102	0.68	.02677	8	.31496	48	1.88976	88	3.46457
0.29	.01142	0.69	.02717	9	.35433	49	1.92913	89	3.50394
0.30	.01181	0.70	.02756	10	.39370	50	1.96850	90	3.54331
0.31	.01220	0.71	.02795	11	.43307	51	2.00787	91	3.58268
0.32	.01260	0.72	.02835	12	.47244	52	2.04724	92	3.62205
0.33	.01299	0.73	.02874	13	.51181	53	2.08661	93	3.66142
0.34	.01339	0.74	.02913	14	.55118	54	2.12598	94	3.70079
0.35	.01378	0.75	.02953	15	.59055	55	2.16535	95	3.74016
0.36	.01417	0.76	.02992	16	.62992	56	2.20472	96	3.77953
0.37	.01457	0.77	.03032	17	.66929	57	2.24409	97	3.81890
0.38	.01496	0.78	.03071	18	.70866	58	2.28346	98	3.85827
0.39	.01535	0.79	.03110	19	.74803	59	2.32283	99	3.89764
0.40	.01575	0.80	.03150	20	.78740	60	2.36220	100	3.93701

Figure 3-3. Equivalents of Metric and English units. (The L. S. Starrett Company)

Vernier Caliper

The vernier caliper (Figure 3-4) is an instrument used for measuring with greater precision than the steel rule. The vernier caliper's two scales are the main scale and the vernier scale. The relationship between the main scale of the beam and the vernier scale of any vernier caliper is:

$$\frac{N}{N + 1}$$

N = number of graduations on the main scale included within a certain length.

$N + 1$ = number of graduations on the vernier scale included within the same length.

Three types of vernier calipers commonly used in machine shop work are the *25-division vernier caliper,* the *50-division vernier caliper*, and the *dual-system vernier caliper.*

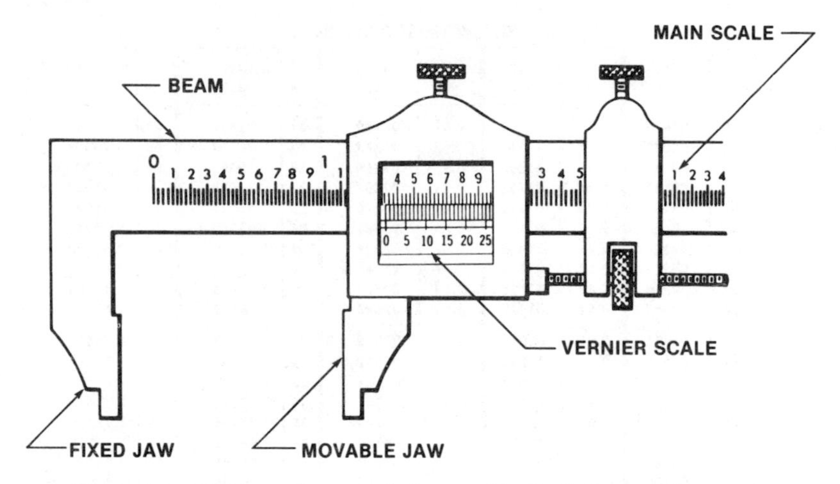

Figure 3-4. Vernier caliper nomenclature. (The L. S. Starrett Company)

25-Division Vernier Caliper. The main scale of the 25-division vernier caliper (Figure 3-4 and 3-5) is graduated in .025″ divisions. Every fourth graduation is marked with the numbers 1 through 9 indicating tenths of an inch (.100″). The vernier scale consists of 25 divisions corresponding with the 24 graduations on the main scale.

The combination of the two scales makes it possible to take measurements with an accuracy of .001″, compared to 1/64″ (.0156″) or 1/100″ (.01″), which is possible with a steel rule. Vernier scale measurements are taken by placing the part to be measured between the two jaws, and noting the reading on both the main scale and the vernier scale. The reading on the main scale is found by taking the closest graduation on the left side of the vernier scale's zero line. The reading on the vernier scale is found by taking the one division of the 25 that aligns with a graduation on the main scale.

In cases where the vernier scale's zero line aligns with a graduation of the main scale, the reading is exactly .025″ or a multiple of .025″ (.050″, .075″, .100″, .200″, and so forth).

Figure 3-5 shows a 25-division vernier scale setting with the vernier scale's zero between 1.425″ and 1.450″, and its eleventh division aligning with a main scale graduation. The total reading in this setting is 1.436″.

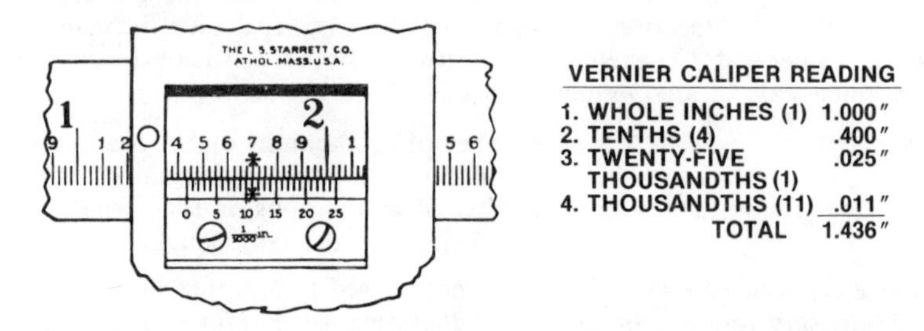

Figure 3-5. A 25-division vernier caliper reading. (The L. S. Starrett Company)

50-Division Vernier Caliper. The main scale on the 50-division vernier caliper is graduated in .050″ divisions. Every second division is marked with numbers 1 through 9, indicating tenths of an inch (.100″). The vernier scale consists of 50 divisions corresponding to the 49 graduations on the main scale.

The combination of the two scales makes it possible to take measurements with an accuracy of .001″. Reading the 50-division vernier caliper is simpler than reading the 25-division vernier caliper because the divisions of the vernier scale are wider, and it is easier to see which division aligns with a graduation on the main scale. The 50-division vernier caliper (Figure 3-6) has two vernier scales—one for inside measurements and one for outside measurements.

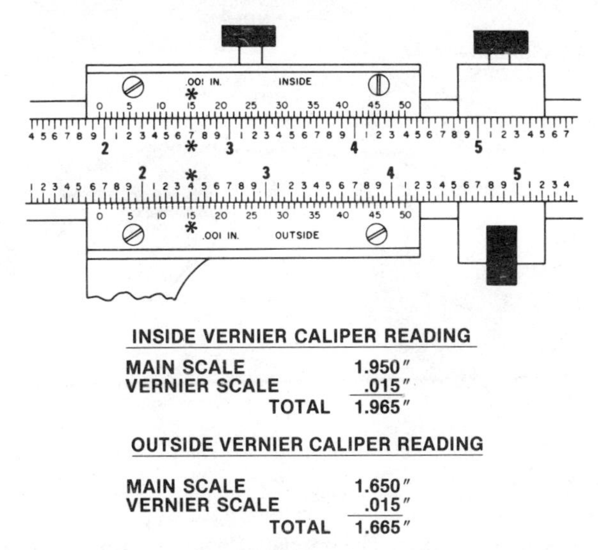

Figure 3-6. A 50-division vernier caliper reading. (The L. S. Starrett Company)

Dual-system Vernier Caliper. The dual-system vernier caliper (Figure 3-7) is graduated in both English and metric units. The vernier scale on the side of the main scale with the inch graduations has 50 divisions, and measures with an accuracy of .001″. The vernier scale on the side with the millimeter graduations has 100 divisions, and measures with an accuracy of 0.02 mm.

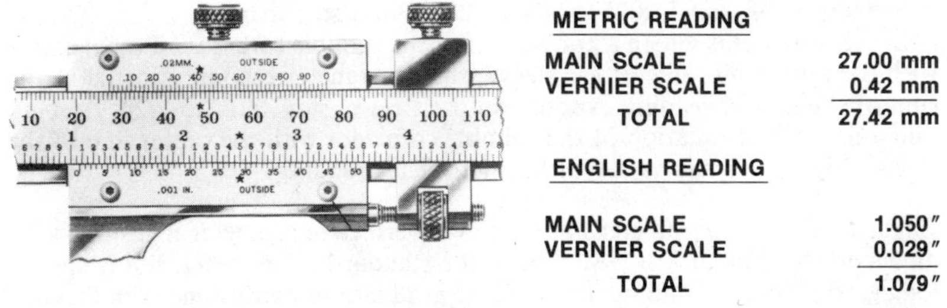

Figure 3-7. A dual-system vernier caliper reading. (The L. S. Starrett Company)

Micrometer

The micrometer (Figure 3-8) is a precision instrument used to measure parts requiring an accuracy of .001″ or .0001″. It consists of an accurately ground screw that rotates in a fixed nut concealed by a sleeve. Micrometer measurements are made between the anvil and the spindle. The anvil corresponds with the fixed jaw, and the spindle corresponds with the movable jaw of a vernier caliper.

Two types of micrometers are the *standard micrometer* and the *vernier micrometer.* Both types are available in English and metric units.

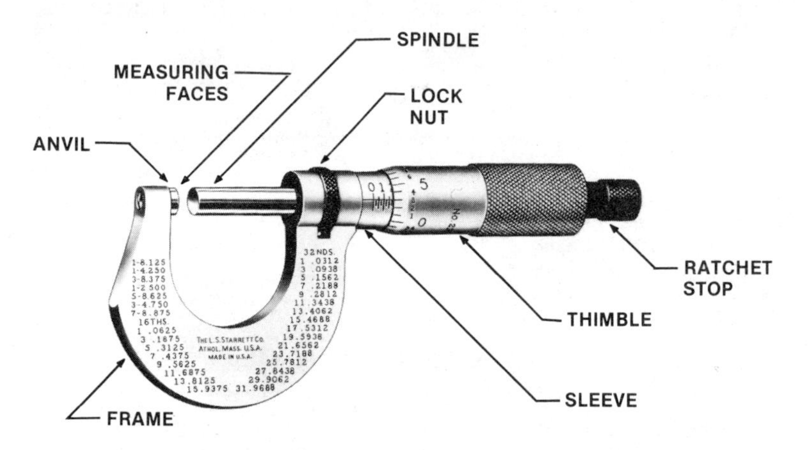

Figure 3-8. Typical micrometer nomenclature. (The L. S. Starrett Company)

Standard English Micrometer. The standard English micrometer measures with an accuracy of .001″. The threaded part of the spindle has 40 threads-per-inch. The cylindrical surface of the sleeve is graduated in forthieths of an inch, therefore, for every turn of the thimble, the spindle moves towards or away from the anvil 1/40″ or .025″. At the same time, the thimble covers, or uncovers one of the graduations on the sleeve. The thimble has 25 divisions marked 0-24. Each of these divisions corresponds to .001″.

When measuring with a micrometer, the largest reading exposed on the sleeve, and one of the 25 graduations on the thimble's beveled edge indicate the length between the anvil and the spindle. When the graduation on the thimble marked with zero aligns with the long center line of the sleeve, the micrometer reading is a multiple of .025″, .075″, .100″, .200″, and so forth.

In measurements where a graduation of the thimble (other than zero) aligns with the long center line of the sleeve, the micrometer reading is equal to the sum of the largest reading exposed on the sleeve, plus .001″ to .024″, depending on which graduation of the thimble coincides with the center line of the sleeve. Figure 3-9 shows a typical micrometer reading.

English Vernier Micrometer. The English vernier micrometer measures with an accuracy of .0001″. It is similar to the standard micrometer, but it has ten long graduations around its sleeve. These graduations correspond with the nine graduations on the thimble.

When measuring with an English vernier micrometer and none of the 25 graduations of the thimble aligns with the center line of the sleeve, the total reading includes a 4-digit decimal number. The number of the fourth digit depends upon which one of the ten long graduations aligns with a graduation of the thimble. Figure 3-10 shows a typical reading.

Figure 3-9. Standard English micrometer reading.

Figure 3-10. English vernier micrometer reading.

Standard Metric Micrometer. The standard metric micrometer measures with an accuracy of 0.01 mm. The threaded part of the spindle has a metric thread with 0.5 mm pitch. Each graduation on the sleeve is 0.5 mm, and each graduation on the thimble is 0.01 mm. For every turn of the thimble, the spindle moves towards or away from the anvil 0.5 mm. The thimble has 50 divisions, therefore, any reading between 0.01 mm and 0.49 mm is indicated by one graduation of the thimble. The total reading of metric micrometer settings is equal to the sum of the reading on the sleeve and the thimble. See Figure 3-11 for a typical reading.

Figure 3-11. Standard metric micrometer reading.

Metric Vernier Micrometer. The metric vernier micrometer measures with an accuracy of 0.002 mm. It is similar to the standard metric micrometer, but it has five long graduations around its sleeve. These graduations correspond with the ten graduations on the thimble.

When measuring with the metric vernier micrometer and none of the 50 graduations of the thimble aligns with the center line of the sleeve, the total reading contains a 3-digit decimal number. The number of the third digit depends upon which one of the five long graduations aligns with a graduation of the thimble. See Figure 3-12 for a typical reading.

Figure 3-12. Metric vernier micrometer reading.

Vernier Bevel Protractor

The vernier bevel protractor (Figure 3-13) is a precision instrument used to measure angles. The vernier scale on the protractor consists of 12 divisions, thus making measurements of $^{1}/_{12}{}^{\circ}$ or 5′ (minutes) possible. The total reading of a vernier protractor is indicated by the number of whole degrees past the vernier's zero line, plus 5′ to 55′. The number of minutes depends upon which of the 12 divisions on the vernier scale coincides with a graduation on the protractor. See Figure 3-14 for a typical reading.

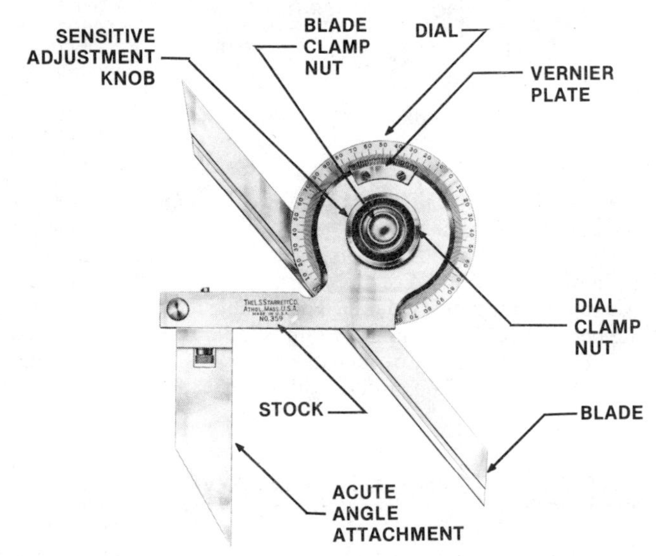

Figure 3-13. Vernier bevel protractor nomenclature. (The L. S. Starrett Company)

Figure 3-14. Vernier bevel protractor reading. (The L. S. Starrett Company)

For additional examples of readings on the vernier bevel protractor and other tools discussed, see Figures 3-15 through 3-22.

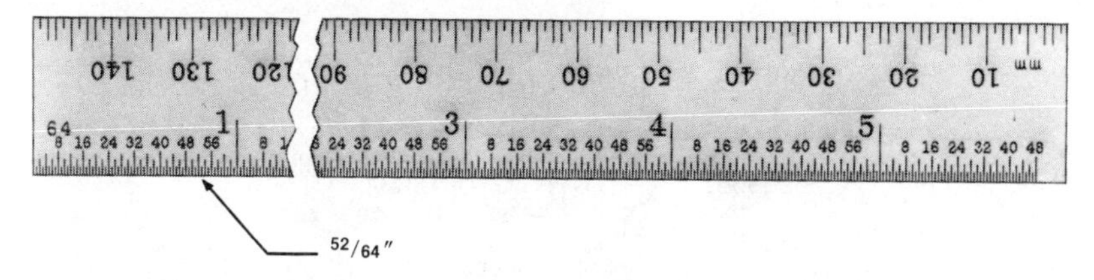

Figure 3-15. English steel rule reading. (The L. S. Starrett Company)

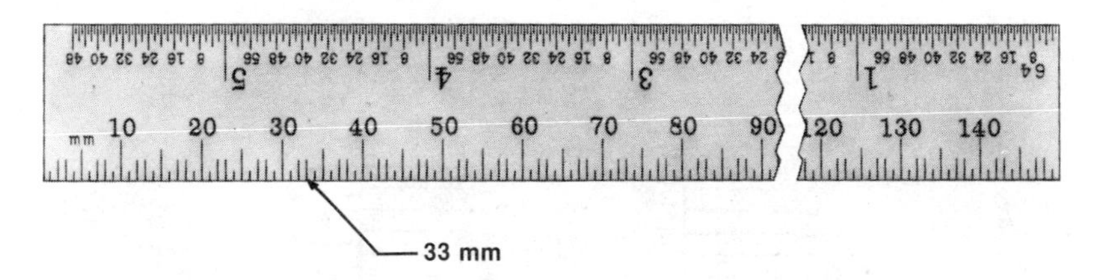

Figure 3-16. Metric steel rule reading. (The L. S. Starrett Company)

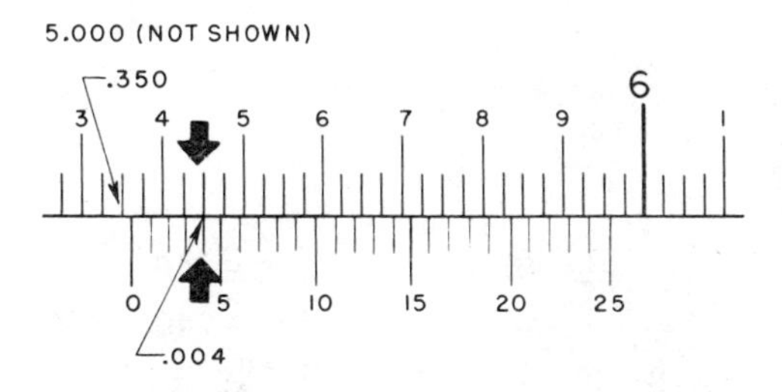

Figure 3-17. A 25-division vernier caliper reading.

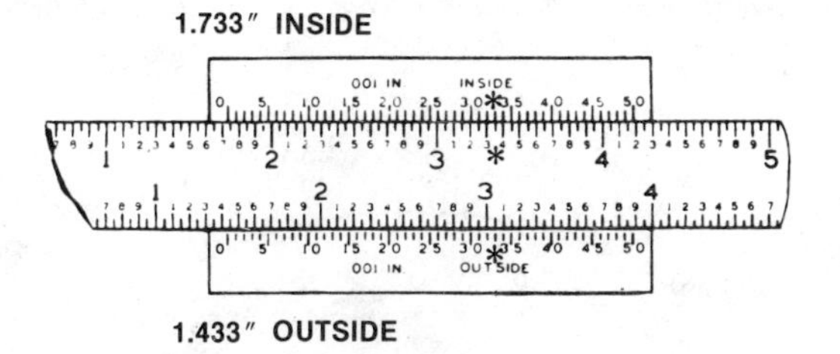

Figure 3-18. A 50-division vernier caliper reading. (The L. S. Starrett Company)

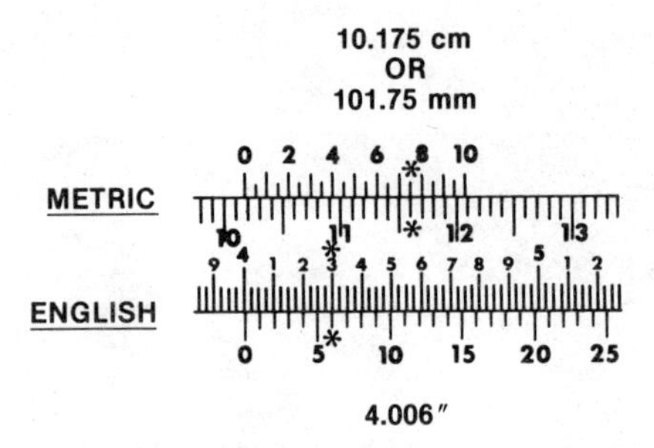

Figure 3-19. A dual-system vernier caliper reading.

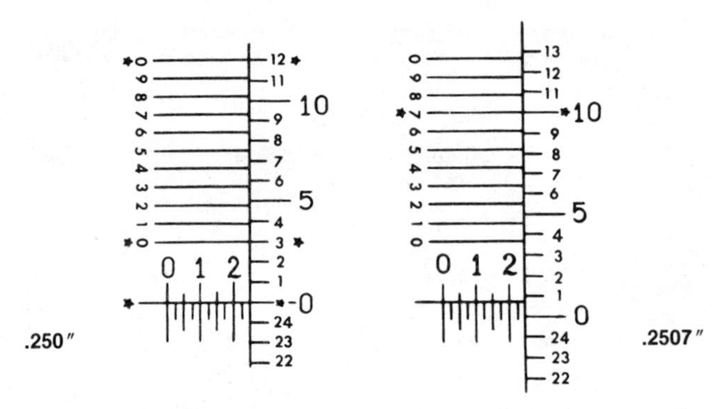

Figure 3-20. English vernier micrometer readings. (The L. S. Starrett Company)

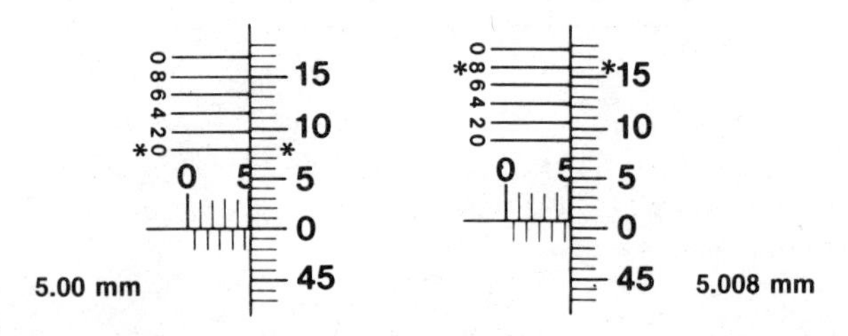

Figure 3-21. Metric vernier micrometer readings. (The L. S. Starrett Company)

Figure 3-22. Universal bevel vernier protractor reading.

COMMON METALS AND ALLOYS

Metals are metallic elements found in the crust of the earth. Alloys are substances composed of two or more elements. The various metals and alloys commonly used in industry are divided into two groups: *ferrous* (containing iron), and *non-ferrous* (not containing iron).

Ferrous Metals and Alloys

Ferrous metals derive from iron and include *cast iron, wrought iron, steel,* and *alloy steel.*

Cast Iron. Cast iron is an alloy of iron and carbon, containing 2.00% to 4.25% carbon. The three types of cast iron are *gray, white,* and *malleable.* Each of these has unique characteristics that make it suitable for certain applications.

Gray and malleable cast irons are softer than white cast iron, and they are used in castings of many machine parts. White cast iron is very hard and brittle and is highly resistant to abrasion. Of the three types of cast iron, the most common and the easiest to machine is gray cast iron.

Wrought Iron. Wrought iron is the purest form of iron because it contains less than 0.05% carbon. Characteristics of wrought iron are its ductility, malleability, weldability, and toughness. Wrought iron is ideal for ornamental iron work and for applications that require hammering or welding.

Steel. Steel is an alloy of iron and carbon containing 0.15% to 1.65% carbon. Depending on its carbon content, steel is classified as *low carbon steel, medium carbon steel,* or *high carbon steel.*

Low carbon steels (0.15% to 0.30% carbon) are used for low-stress applications and for machine parts that do not need hardening. Low carbon steels cannot be hardened.

Medium carbon steels (0.30% to 0.60% carbon) are used in applications where high tensile stress is necessary. They are suitable for forgings and tools such as hammers, wrenches, and screwdrivers. Medium carbon steels can be hardened.

High carbon steels (0.60% to 1.65% carbon) are used in applications that require strength, hardness, toughness, and wear resistance. They are suitable for a variety of machine parts that require high strength, and for cutting tools such as drills, chisels, and punches. High carbon steels can be hardened, and they have higher degrees of hardness than medium carbon steels.

Alloy steel. Alloy steel is a carbon steel containing one or more additional elements in small quantities. Among the most common elements added to carbon steel to make alloy steels are *chromium, manganese, molybdenum, nickel, silicon, tungsten,* and *vanadium.* Adding small quantities of one or more of these alloying elements to carbon steel changes its chemical and physical properties. This makes the alloy steel suitable for particular uses.

Classification of Steels

Two coding systems exist for classifying carbon steels and alloy steels. The first is sponsored by the *American Iron and Steel Institute* (AISI), and the second is sponsored by the *Society of Automotive Engineers* (SAE). Both of these systems use a 4 or 5-digit number to identify the major alloying element or elements, and the carbon content in a particular steel.

In addition to the 4 or 5-digit numbers, the AISI system also uses a letter prefix to indicate the method by which the particular steel was produced. The code for the prefix letters and the first digit numbers are:

Letters Identifying Production Method	Numbers Identifying Major Alloying Elements	
B - Acid bessemer carbon steel	1. Carbon	6. Chromium-vanadium
C - Basic open-hearth carbon steel	2. Nickel	7. Tungsten
D - Acid open-hearth	3. Nickel-chromium	8. Nickel-chromium-molybdenum
E - Electric furnace	4. Molybdenum	
	5. Chromium	9. Silicon-manganese

Figure 3-23 lists the AISI-SAE code classification system for carbon and alloy steels.

In both systems, the first digit indicates the major alloying element; the second digit indicates the approximate percentages of the major and the other alloying elements; and the last two digits indicate the percentage of carbon content.

TYPE OF STEEL	SERIES DESIGNATION
Carbon steels	1XXX
Plain carbon	10XX
Free-machining, resulfurized (screw stock)	11XX
Free-machining, resulfurized, rephosphorized	12XX
Manganese steels	13XX
High-manganese carburizing steels	15XX
Nickel steels	2XXX
3.50% nickel	23XX
5.00% nickel	25XX
Nickel-chromium steels	3XXX
1.25% nickel, 0.60% chromium	31XX
1.75% nickel, 1.00% chromium	32XX
3.50% nickel, 1.50% chromium	33XX
Corrosion-and heat-resisting steels	30XXX
Molybdenum steels	4XXX
Carbon-molybdenum	40XX
Chromium-molybdenum	41XX
Chromium-nickel-molybdenum	43XX
Nickel-molybdenum	46XX and 48XX
Chromium steels	5XXX
Low chromium	51XX
Medium chromium	52XXX
Corrosion-and heat resisting	51XXX
Chromium-vanadium steels	6XXX
Chromium 1.0%	61XX
Nickel-chromium-molybdenum	86XX and 87XX
Manganese-silicon	92XX
Nickel-chromium-molybdenum	93XX
Manganese-nickel-chromium-molybdenum	94XX
Nickel-chromium-molybdenum	97XX
Nickel-chromium-molybdenum	98XX
Boron (0.0005% boron minimum)	XXBXX

Figure 3-23. AISI-SAE code classification system for classifying steels and alloy steels.

For example, a particular steel specified as 2318 indicates the following:
1. First-digit number 2: nickel alloy
2. Second-digit number 3: approximately 3.5% nickel content
3. Third and fourth-digit numbers 18: .18% carbon content

Hardening of Steels

Steels containing .30% or more carbon can be hardened. The purpose of hardening steels is to increase their strength and their resistance to scratching, cutting, and abrading. Many methods exist for hardening steels. Some of these methods increase the hardness of the whole part, while other methods increase the hardness of the part's surface only.

The hardness of material is determined by testing its relative resistance to a penetrator of a known shape and size. The hardness is identified by a number, and the number depends on the type of test conducted. Therefore, the hardness of a specific material can be identified by different numbers (Figure 3-24).

Non-ferrous Metals

Non-ferrous metals are metals that contain no iron. One of the main characteristics of non-ferrous metals is their resistance to corrosion. The most common non-ferrous metals used in manufacturing are *aluminum, copper, lead, magnesium, nickel, tin,* and *zinc.*

Aluminum. Aluminum is a light, soft metal that is whitish in color. It has very good machinability, weldability, electrical conductivity, and resistance to corrosion. Aluminum is used either as a pure metal (99.0% pure) or as a base metal in several alloys, among which the most common are aluminum-silicon, aluminum-magnesium, aluminum-copper-silicon, aluminum-silicon-magnesium, and aluminum-zinc-magnesium. Aluminum or its alloys are available in almost all commercial forms in which metals are commonly used.

Copper. Copper is a heavy, soft, reddish-colored metal, and is very malleable. It has excellent conductivity and corrosion resistance. Copper, like aluminum, is used either as pure metal (99.0% pure) or as a base metal in several alloys. Copper and its alloys are used in applications where color, ease of forming or joining, conductivity, strength, and/or corrosion resistance is desirable.

Two commonly used copper alloys are the copper-zinc alloy and the copper-tin alloy. The copper-zinc alloy is known as *brass.* This alloy is used extensively in manufacturing. The copper-tin alloy is known as *bronze.* This alloy is harder than brass and is used as material for bearings.

Lead. Lead is very heavy and soft. It is a silvery, bright-colored metal when cut, but it turns gray when exposed to air. It is also very malleable and has high corrosion resistance. Lead is used either as pure metal for the manufacture of pipes, conduits, and linings, or it is used as an alloy. Two common lead alloys are lead-tin and lead-tin-antimony. Both of these alloys are used for soldering.

Magnesium. Magnesium is the lightest structural metal. It has high strength in relation to its weight, good machinability, and very good corrosion resistance. Magnesium is used mainly as an alloy with aluminum and manganese in sand and permanent mold castings, die castings, sheets and plates, extruded forms,

Rockwell C-scale Hardness Number†	Diamond Pyramid Hardness Number, Vickers	BRINELL HARDNESS NUMBER 10 mm BALL, 3,000-kg LOAD			ROCKWELL HARDNESS NUMBER†			Shore Scleroscope Hardness Number	Tensile Strength (approximate) in 1,000 psi	Rockwell C-scale Hardness Number†
		Standard ball	Hultgren penetrator	Tungsten carbide ball	A-scale 60-kg-load, brale penetrator	B-scale 100-kg load, 1/16" diameter ball	D-scale 100-kg load, brale penetrator			
68	940	....	....	....	85.6	....	76.9	97	....	68
67	900	....	....	....	85.0	....	76.1	95	....	67
66	865	....	....	....	84.5	....	75.4	92	....	66
65	832	....	....	739	83.9	....	74.5	91	....	65
64	800	....	....	722	83.4	....	73.8	88	....	64
63	772	....	....	705	82.8	....	73.0	87	....	63
62	746	....	....	688	82.3	....	72.2	85	....	62
61	720	....	....	670	81.8	....	71.5	83	....	61
60	697	....	613	654	81.2	....	70.7	81	....	60
59	674	....	599	634	80.7	....	69.9	80	326	59
58	653	....	587	615	80.1	....	69.2	78	315	58
57	633	....	575	595	79.6	....	68.5	76	305	57
56	613	....	561	577	79.0	....	67.7	75	295	56
55	595	....	546	560	78.5	....	66.9	74	287	55
54	577	....	534	543	78.0	....	66.1	72	278	54
53	560	....	519	525	77.4	....	65.4	71	269	53
52	544	500	508	512	76.8	....	64.6	69	262	52
51	528	487	494	496	76.3	....	63.8	68	253	51
50	513	475	481	481	75.9	....	63.1	67	245	50
49	498	464	469	469	75.2	....	62.1	66	239	49
48	484	451	455	455	74.7	....	61.4	64	232	48

Figure 3-24. Typical hardness tests and hardness number conversions. (Butterfield)

and forgings. Magnesium alloys are used extensively in the aircraft and aerospace industries.

Nickel. Nickel is a grayish-white metal that resists corrosion, and is capable of taking a bright polish. Nickel is used as an alloying element in steel alloys, including stainless steels, and it is used as the base metal in several nickel alloys such as *monel metal, hastelloy,* and *inconel.*

Monel metal contains nickel, copper, and small amounts of manganese or aluminum. It is used in valve seats, marine pumps, and non-magnetic aircraft parts. Hastelloy contains nickel, silicon, and copper. It is used extensively in the chemical industry. Inconel contains nickel, chromium, and iron. It is used in food processing equipment, aircraft exhaust mainfolds, and heat-treating furnaces and equipment.

Tin. Tin is a silvery, whitish metal. It is used mainly as an alloying element in copper-tin alloys and in lead-tin alloys (*solders*). It is also used as the base metal in *pewter*, a tin alloy containing tin, antimony, and copper. Pewter is used extensively in the housewares industry.

Zinc. Zinc is a relatively strong metal characterized by toughness and excellent corrosion resistance. It is used as a plating material for galvanized coatings, as a base metal in zinc alloys, and as an alloying element in copper-zinc alloys. Zinc alloys are used extensively in the automotive industry and in general manufacturing industries.

NOTE: Figure 3-25 summarizes the physical characteristics of selected metals and alloys.

Identification of Steels and Other Metals

Metals are usually identified by the *color code method* and the *spark-test method*.

Color Code Method. The color code method identifies a metal, or alloy bar or rod, by the color that is painted on the end of it. Figure 3-26 shows the color code used for steel identification.

In laboratories and shops where many different types of metal stock are used, it is a good practice to store them separately. To avoid mistakes, use the painted side of the stock only after all the rest is used.

Spark-test Method. The spark-test method identifies a steel or alloy by grinding the steel or alloy, and comparing the spark patterns produced with known spark patterns of similar alloy steels (Figure 3-27).

METAL SAWING

Metal sawing is the cutting of metal stock using a thin metal blade with sharp teeth on one edge. Sawing may be done with a hand hacksaw, power hacksaw, or band saw.

Saw Blades

Saw blades are made of carbon steel or alloy steel. They can cut any common metal or alloy except hardened steel. See Figure 3-28 for blade characteristics and terminology.

METAL OR COMPOSITION	CHEMICAL SYMBOL	SPECIFIC GRAVITY	WEIGHT PER CUBIC INCH, POUND	WEIGHT PER CUBIC FOOT, POUNDS	MELTING POINT DEG. F	STRUCTURE	LINEAR EXPANSION PER UNIT LENGTH PER DEG. F	ELECTRIC CONDUCTIVITY; SILVER = 100
Aluminum	Al	2.56	0.0924	159.7	1218	*M*	0.00001233	63.00
Antimony	Sb	6.71	0.2422	418.7	1166	*B*	0.00000587	3.59
Barium	Ba	3.75	0.1354	234.0	1562	*M*	....	30.61
Bismuth	Bi	9.80	0.3538	611.5	520	*B*	0.00000731	1.40
Boron	B	2.50	0.0939	162.2	4,000 - 4,500	*H*	....	....
Brass: 80 C., 20 Z	....	8.60	0.3105	536.6				
70 C., 30 Z	....	8.40	0.3032	524.1	1,700 - 1,850	*M*	0.00001042	....
60 C., 40 Z	....	8.35	0.3018	521.7				
50 C., 50 Z	....	8.20	0.2960	511.6				
Bronze	....	8.35	0.3195	552.2	1675	*B*	0.00001024	....
Cadmium	Cd	8.60	0.3105	536.6	610	*M*	0.00001755	24.38
Calcium	Ca	1.57	0.0567	98.0	1490	*M*	....	21.77
Chromium	Cr	6.50	0.2347	405.6	2939	*B*	....	16.00
Cobalt	Co	8.65	0.3123	539.8	2696	*M*	0.00000687	16.93
Copper	Cu	8.82	0.3184	550.4	1981	*M*	0.00000926	97.67
Gold	Au	19.32	0.6975	1205.6	1945	*M*	0.00000817	76.71
Iridium	Ir	22.42	0.8094	1399.0	4260	*M*	0.00000356	13.52
Iron, cast	Fe	7.20	0.2600	449.2	2300	*B*	0.00000589	....
Iron, wrought	Fe	7.85	0.2834	489.8	2750	*M*	0.00000648	16.80
Lead	Pb	11.37	0.4105	709.5	621	*S*	0.00001505	8.42
Magnesium	Mg	1.74	0.0628	103.6	1204	*M*	0.00001497	39.44
Manganese	Mn	7.42	0.2679	463.0	2246	*B*	....	15.75

continued

METAL OR COMPOSITION	CHEMICAL SYMBOL	SPECIFIC GRAVITY	WEIGHT PER CUBIC INCH, POUND	WEIGHT PER CUBIC FOOT, POUNDS	MELTING POINT DEG. F	STRUCTURE	LINEAR EXPANSION PER UNIT LENGTH PER DEG. F	ELECTRIC CONDUCTIVITY; SILVER = 100
Mercury 60° F	Hg	13.58	0.4902	847.4	—38	*F*	....	1.75
Molybdenum	Mo	8.56	0.3090	534.2	4620	*B*	....	17.60
Nickel	Ni	8.80	0.3177	549.1	2646	*M*	0.00000710	12.89
Platinum, rolled	Pt	22.67	0.8184	1414.6	3191	*M*	0.00000499	14.43
Platinum, wire	Pt	21.04	0.7595	1312.9				
Potassium	K	0.87	0.0314	54.3	144	*S*	0.00004611	19.62
Silver	Ag	10.53	0.3802	657.1	1761	*M*	0.00001067	100.00
Sodium	Na	0.98	0.0354	61.1	207	*S*	....	31.98
Steel	Fe	7.80	0.2816	486.7	2500	*M*	0.00000638	12.00
Tellurium	Te	6.25	0.2256	390.0	846	*B*	0.00002048	0.001
Tin	Sn	7.29	0.2632	454.8	449	*M*	0.00001276	14.39
Titanium	Ti	3.54	0.1278	220.9	3272	*M*	....	13.73
Tungsten	W	18.77	0.6776	1171.2	6152	*B*	....	14.00
Vanadium	Va	5.50	0.1986	343.2	3128	*M*	....	4.95
Zinc, cast	Zn	6.86	0.2476	428.1	787	*B*	0.00001653	29.57
Zinc, rolled	Zn	7.15	0.2581	446.1		*M*		

B = **brittle;** *F* = **fluid;** *H* = **hard;** *M* = **malleable;** *S* = **soft.**

Figure 3-25. Physical characteristics of selected metals and alloys. (Dietzgen Corporation)

SAE NUMBER	CODE COLOR
CARBON STEELS	
1010.	White
1015	White
X1015	White
1020	Brown
X1020	Brown
1025	Red
X1025	Red
1030	Blue
1035	Blue
1040	Green
X1040	Green
1045	Orange
X1045	Orange
1050	Bronze
1095	Aluminum
FREE-CUTTING STEELS	
1112	Yellow
X1112	Yellow
1120	Yellow and brown
X1314	Yellow and blue
X1315	Yellow and red
X1335	Yellow and black
NICKEL STEELS	
2015	Red and brown
2115	Red and bronze
2315	Red and blue
2320	Red and blue
2330	Red and white
2335	Red and white
2340	Red and green
2345	Red and green
2350	Red and aluminum
2515	Red and black
MOLYBDENUM STEELS	
4130	Green and white
X4130	Green and bronze
4135	Green and yellow
4140	Green and brown
4150	Green and brown
4340	Green and aluminum
4345	Green and aluminum
4615	Green and black
4620	Green and black
4640	Green and pink
4815	Green and purple
X1340	Yellow and black
MANGANESE STEELS	
T1330	Orange and green
T1335	Orange and green
T1340	Orange and green
T1345	Orange and red
T1350	Orange and red
NICKEL - CHROMIUM STEELS	
3115	Blue and black
3120	Blue and black
3125	Pink
3130	Blue and green
3135	Blue and green
3140	Blue and white
X3140	Blue and white
3145	Blue and white
3150	Blue and brown
3215	Blue and purple
3220	Blue and purple
3230	Blue and purple
3240	Blue and aluminum
3245	Blue and aluminum
3250	Blue and bronze
3312	Orange and black
3325	Orange and black
3335	Blue and orange
3340	Blue and orange
3415	Blue and pink
3435	Orange and aluminum
3450	Black and bronze
4820	Green and purple
CHROMIUM STEELS	
5120	Black
5140	Black and white
5150	Black and white
52100	Black and brown
CHROMIUM-VANADIUM STEELS	
6115	White and brown
6120	White and brown
6125	White and aluminum
6130	White and yellow
6135	White and yellow
6140	White and bronze
6145	White and orange
6150	White and orange
6195	White and purple
TUNGSTEN STEELS	
71360	Brown and orange
71660	Brown and bronze
7260	Brown and aluminum
SILICON - MANGANESE	
9255	Bronze and aluminum
9260	Bronze and aluminum

Figure 3-26. Partial color code for steel identification.

One of the most important factors for efficient metal sawing is to use a saw blade with the proper *pitch* (number of teeth-per-inch). When choosing the pitch of a saw blade for a particular job, the hardness and thickness of the metal to be cut must be considered. As a rule, hard and thin metals require finer pitch than soft and thick metals. The number of teeth-per-inch of common hand saw blades is 14T, 18T, 24T, and 32T. For power hacksaw and band saw blades, the number of teeth-per-inch is 3T, 4T, 6T, 8T, 10T, and 14T.

When choosing a saw blade for a particular job, make sure that at least two teeth are in contact with the stock. See Figures 3-29, 3-30, and 3-31 for information related to saw blade selection and sawing speed.

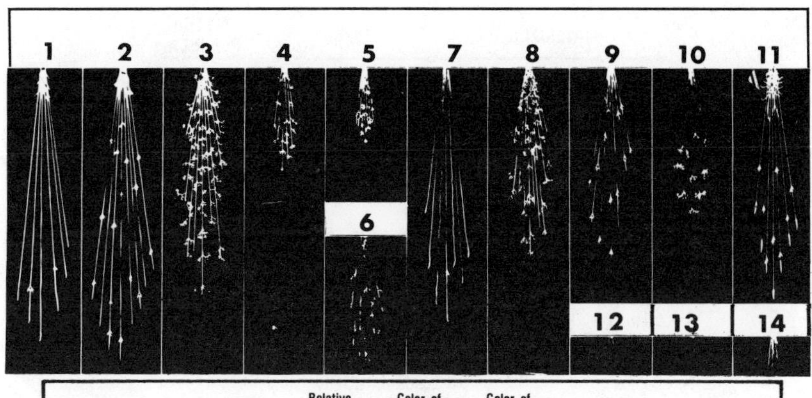

Metal	Volume of Stream	Relative Length of Stream, Inches†	Color of Stream Close to Wheel	Color of Streaks, Near End of Stream	Quantity of Spurts	Nature of Spurts
1. Wrought iron	Large	65	Straw	White	Very few	Forked
2. Machine steel (AISI 1020)	Large	70	White	White	Few	Forked
3. Carbon tool steel	Moderately large	55	White	White	Very many	Fine, repeating
4. Gray cast iron	Small	25	Red	Straw	Many	Fine, repeating
5. White cast iron	Very small	20	Red	Straw	Few	Fine, repeating
6. Annealed mall iron	Moderate	30	Red	Straw	Many	Fine, repeating
7. High-speed steel (18-4-1)	Small	60	Red	Straw	Extremely few	Forked
8. Austenitic manganese steel	Moderately large	45	White	White	Many	Fine, repeating
9. Stainless steel (Type 410)	Moderate	50	Straw	White	Moderate	Forked
10. Tungsten-chromium die steel	Small	35	Red	Straw*	Many	Fine, repeating*
11. Nitrided Nitralloy	Large (curved)	55	White	White	Moderate	Forked
12. Stellite	Very small	10	Orange	Orange	None	
13. Cemented tungsten carbide	Extremely small	2	Light orange	Light orange	None	
14. Nickel	Very small**	10	Orange	Orange	None	
– Copper, brass, aluminum	None				None	

†Figures obtained with 12″ wheel on beach sand and are relative only. Actual length in each instance will vary with grinding wheel, pressure, etc. *Blue-white spurts. **Some wavy streaks

Figure 3-27. Spark test patterns and colors in grinding metals. (Norton Company)

Figure 3-28. Pitch is the number of teeth-per-inch.

MATERIAL	TYPE	TEETH-PER-INCH
Aluminum	Solids	14
Angles	Heavy	18
Angles	Light	24
Babbitt		14
Brass	Solids up to 1″	18
Brass plpe		24
Brass tubing		24
Bronze	Solids up to 1″	18
BX cable	Heavy	24
BX cable	Light	32
Cast iron	Up to 1″	18
Channel	Heavy	18
Channel	Light	24
Cable	Heavy	18
Copper	Solids up to 1″	14
Drill rod	Over 1/4″	18
Drill rod	No. 30 to 1/4″	24
Drill rod	No. 30 and smaller	32
General purpose cutting		18
Iron pipe		24
Metal conduit		24
Sheet metal	Over 18 gauge	24
Sheet metal	Under 18 gauge	32
Steels	1/4″ to 1″	18
Steels	1/4″ and under	24
Tubing	Over 18 gauge	24
Tubing	Under 18 gauge	32

Figure 3-29. Recommendations for the use of hacksaw blades on different types of material. (The L. S. Starrett Company)

MATERIAL	TEETH-PER-INCH	STROKES-PER-MINUTE	FEED PRESSURE
Aluminum alloy	4-6	150	Light
Aluminum, pure	4-6	150	Light
Brass castings, soft	6-10	150	Light
Brass castings, hard	6-10	135	Light
Bronze castings	6-10	135	Medium
Cast iron	6-10	135	Medium
Copper, drawn	6-10	135	Medium
*Carbon tool steel	6-10	90	Medium
*Cold rolled steel	4-6	135	Heavy
*Drill rod	10	90	Medium
*High-speed steel	6-10	90	Medium
*Machinery steel	4-6	135	Heavy
Manganese bronze	6-10	90	Light
*Malleable iron	6-10	90	Medium
*Nickel silver	6-10	60	Heavy
*Nickel steel	6-10	90	Heavy
Pipe, iron	10-14	135	Medium
Slate	6-10	90	Medium
*Structural steel	6-10	135	Medium
Tubing, brass	14	135	Light
*Tubing, steel	14	135	Light

* Use cutting compounds or coolant.

Figure 3-30. Power hacksaw blade recommendations for use on various types of material. (The L. S. Starrett Company)

MATERIAL	SIZE OF MATERIAL			SIZE OF MATERIAL		
	½″ - 1″	1″ - 2″	Over 2″	½″ - 1″	1″ - 2″	Over 2″
	TEETH-PER-INCH			SPEED FPM		
STEEL SAE						
Carbon steel #1010 - #1095	14	10	6	175	150	125
Free-machining #X1112 - #X1340	12	8	6	250	200	150
Nickel chrome #2115 - #3415	14	10	6	100	85	60
Molybdenum #4023 - #4820	14	10	6	125	100	75
Chromium #5120 - #52100	14	10	8	100	75	50
Tungsten #7260 - #71360	14	10	8	85	60	50
N.E. steels #8024 - #8949	14	10	6	175	150	100
Silicon manganese #9255 - #9260	14	10	6	100	75	50
STEELS MISC.						
Armor plate	14	12	6	100	75	50
Graphic steel	14	12	6	150	125	75
High-speed steel	14	10	8	100	75	50
Stainless steel	12	10	8	60	50	40
Angle iron	14	14	10	190	175	150
Pipe	14	12	8	250	225	185
I Beams and channels	14	14	10	250	200	175
Tubing (thinwall)	14	14	14	250	200	200
Cast steels	14	12	8	150	75	50
Cast iron	12	10	8	200	185	160
NON-FERROUS METALS						
Aluminum (all types)	8	6	6	250	250	250
Brass	8	8	8	250	250	250
Bronze (cast)	10	8	8	185	125	50
Bronze (rolled)	12	10	6	175	125	75
Beryllium	10	8	6	175	150	125
Copper	10	8	6	250	225	225
Magnesium	8	8	6	250	250	250
Kirksite	10	8	6	200	175	150
Monel metal	10	8	6	100	75	50
Zinc	8	6	6	250	225	200
NON-METALS						
Bakelite	10	8	6	250	250	250
Carbon	10	8	6	250	250	250
Plastics (all types)	12	8	8	250	250	250
Wood	8	8	6	250	250	250

Figure 3-31. Recommendations (teeth-per-inch and speed) for band saw blades used on different types of material. (The L. S. Starrett Company)

PRINCIPLES OF MACHINING PROCESSES

Machining is a process in which material is removed from a workpiece (stock) to make it conform to specified dimensions and a specified shape. The basic types of machining processes are drilling, shaping or planing, turning, boring, milling, and grinding.

Machining processes involve the relative movements of the cutting tool and the workpiece. These movements are the speed (V), feed (f), and depth-of-cut. See Figure 3-32.

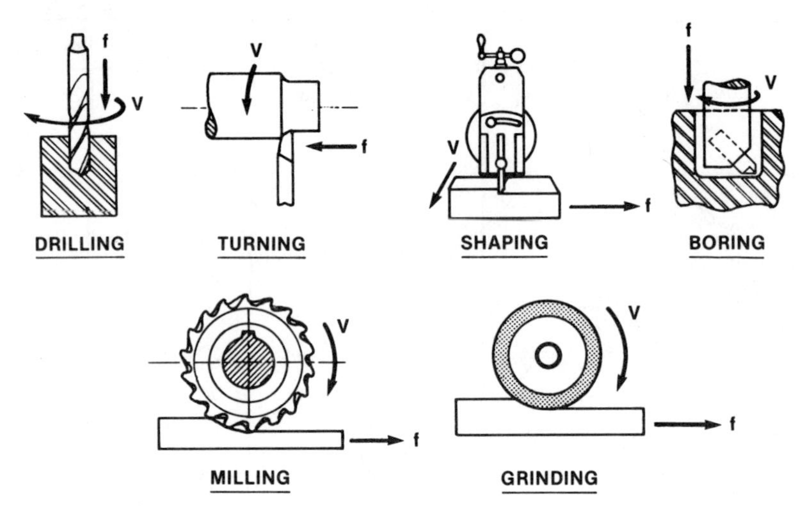

Figure 3-32. Basic machining processes are based on the relative movements of the cutting tool and workpiece.

Conditions for Material Removal

In addition to the speed, feed, and depth-of-cut, to remove material from a workpiece, the following conditions must exist:

1. The cutting-tool material must be harder than the workpiece material.
2. The cutting tool must be properly ground with a sharp edge, or edges, and it must be mounted rigidly.
3. The workpiece must be mounted rigidly on its holding device.
4. The drive mechanism of the machine tool must have sufficient power to overcome the resisting force during the cutting operation in accordance with the cutting speed, feed, and depth-of-cut.

Machinability

Machinability refers to the relative ease with which a material can be machined. Machinability involves factors such as tool life, cutting speed, feed, depth-of-cut, cutting tool design, cutting fluid, and most importantly, the hardness of the material being cut.

A *machinability rating number* shows the relative ease or difficulty with which various materials are machined. For free-cutting steel, SAE 1112 or AISI B1112, the machinability number is 100. Materials with machinability ratings smaller than 100 are more difficult to machine than free-cutting steel. Likewise, materials with machinability ratings greater than 100 are easier to machine than free-cutting steel. Figure 3-33 shows the relationship between strength, hardness, and machinability ratings of various metals and alloys.

In every machining operation, the recommended *cutting speed* (V), *feed* (f), and *depth-of-cut* (d) have a direct relationship to the machinability rating of the material being machined. The values of these three factors determine the surface quality and the time required for a machining operation.

SAE NUMBER	AISI NUMBER	TENSILE STRENGTH (psi)	HARDNESS (BRINELL)	MACHIN-ABILITY RATING (percent)
Carbon Steels				
1015	C1015	65,000	137	50
1020	C1020	67,000	137	52
X1020	C1022	69,000	143	62
1025	C1025	70,000	130	58
1030	C1030	75,000	138	60
1035	C1035	88,000	175	60
1040	C1040	93,000	190	60
1045	C1045	99,000	200	55
1095	C1095	100,000	201	45
Free-cutting Steels				
X1113	B1113	83,000	193	120 - 140
1112	B1112	67,000	140	100
....	C1120	69,000	117	80
Manganese Steels				
X1314	....	71,000	135	94
X1335	A1335	95,000	185	70
Nickel Steels				
2315	A2317	85,000	163	50
2330	A2330	98,000	207	45
2340	A2340	110,000	225	40
2345	A2345	108,000	235	50
Nickel-Chromium Steels				
3120	A3120	75,000	151	50
3130	A3130	100,000	212	45
3140	A3140	96,000	195	57
3150	A3150	104,000	229	50
3250	....	107,000	217	44
Molybdenum Steels				
4119	....	91,000	179	60
X4130	A4130	89,000	179	58
Molybdenum Steels (continued)				
4140	A4140	90,000	187	56
4150	A4150	105,000	220	54
X4340	A4340	115,000	235	58
4615	A4615	82,000	167	58
4640	A4640	100,000	201	69
4815	A4815	105,000	212	55
Chromium Steels				
5120	A5120	73,000	143	50
5140	A5140		174 - 229	60
52100	E52101	109,000	235	45
Chromium-Vanadium Steels				
6120	A6120		179 - 217	50
6150	A6150	103,000	217	50
Other Alloys and Metals				
Aluminum (11S)		49,000	95	300 - 2000
Brass, Leaded		55,000	RF 100	150 - 600
Brass, Red or Yellow		25 - 35,000	40 - 55	200
Bronze, Lead-bearing		22 - 32,000	30 - 65	200 - 500
Cast Iron, Hard		45,000	220 - 240	50
Cast Iron, Medium		40,000	193 - 220	65
Cast Iron, Soft		30,000	160 - 193	80
Cast Steel (0.35 C)		86,000	170 - 212	70
Copper (F.M.)		35,000	RF 85	65
Low-Alloy, High-Strength Steel		98,000	187	80
Magnesium Alloys				500 - 2,000
Malleable Iron				
Standard		53 - 60,000	110 - 145	120
Pearlitic		80,000	180 - 200	90
Pearlitic		97,000	227	80
Stainless Steel (12% Cr F.M.)		120,000	207	70

Figure 3-33. Relationship between tensile strength, hardness, and machinability ratings of common steels and other metals and alloys.

Cutting Speed. Cutting speed is the peripheral, or surface speed of a rotating cutting tool such as a drill or milling cutter or the surface speed of the workpiece in a lathe. The cutting speed (V), also abbreviated CS, is expressed in feet-per-minute (ft/min or FPM). When using metric units, the cutting speed is expressed in meters-per-minute (m/min). See Figure 3-34 for recommended cutting speeds for selected materials and processes.

Feed. Feed is the rate at which the cutting tool advances at each revolution of the machine tool spindle. It is typically expressed in inches-per-revolution (ipr) or millimeters-per-revolution (mmpr) for drilling, turning, and boring; in inches-per-stroke (in/st) or millimeters-per-stroke (mm/st) for shaping and planing; and in inches-per-tooth (ipt) or millimeters-per-tooth (mmpt) for milling. Feed for each of these operations is covered in more detail later in this chapter under MACHINE TOOLS.

Depth-of-Cut. Depth-of-cut is the setting of penetration of the cutting tool for each cutting (pass). The depth-of-cut is expressed in inches or millimeters (mm).

MATERIAL	FOR HSS CUTTING TOOLS							
	DRILL PRESS		LATHE		SHAPER-PLANER		MILLING MACHINE	
	ft/min	m/min	ft/min	m/min	ft/min	m/min	ft/min	m/min
Machine steel	65 - 90	20 - 28	90 - 100	27 - 30	80 - 100	24 - 30	70 - 100	21 - 30
Tool steel	40 - 65	12 - 20	70 - 90	21 - 27	50 - 60	15 - 18	60 - 70	18 - 20
Cast iron	50 - 80	15 - 25	60 - 80	18 - 24	60 - 70	18 - 21	50 - 80	15 - 24
Bronze	up to 200	up to 60	90 - 100	27 - 30	up to 200	up to 60	65 - 120	20 - 35
Aluminum	up to 400	up to 120	200 - 300	61 - 93	up to 200	up to 60	up to 1,000	up to 300

NOTE: CUTTING SPEEDS IN m/min ARE THE APPROXIMATE EQUIVALENTS IN ft/min.

Figure 3-34. Recommended cutting speed for selected materials and processes.

Spindle Speed

Spindle speed is the speed of a machine tool spindle. It is expressed in number of revolutions-per-minute (N), and is identified by the symbols RPM (English) or U/min (Metric). Spindle speed is related to cutting speed as shown:

$$\text{Spindle Speed} = \frac{\text{Cutting Speed or Surface Speed}}{\text{Circumference of the Rotating Tool or Workpiece}}$$

The following are formulas used to calculate spindle speeds:

ENGLISH UNITS AND SYMBOLS	METRIC UNITS AND SYMBOLS
$N = \frac{12 \times CS}{\pi \times D} = RPM$	$N = \frac{1{,}000 \times V}{\pi \times D} = U/min$
N = number of revolutions-per-minute	N = number of revolutions-per-minute
CS = cutting speed in ft/min	V = cutting speed in m/min
D = diameter of rotating tool or workpiece in inches	D = diameter of rotating tool or workpiece in millimeters (mm)
π = 3.14	π = 3.14
12 = constant for converting feet to inches	1,000 = constant for converting meters to millimeters

See Figures 3-35 and 3-36, and Examples 3-1 and 3-2.

Diameter in inches	CUTTING SPEED IN ft/min												
	35′	40′	45′	50′	60′	70′	80′	90′	100′	110′	120′	130′	140′
	Revolutions-per-minute												
1/16	2139	2445	2750	3056	3667	4278	4889	5500	6112	6723	7333	7945	8556
1/8	1070	1222	1375	1528	1833	2139	2445	2750	3056	3361	3667	3973	4278
3/16	713	815	917	1019	1222	1426	1630	1833	2037	2241	2445	2648	2852
1/4	535	611	688	764	917	1070	1222	1375	1528	1681	1833	1986	2139
5/16	428	489	550	611	733	856	978	1100	1222	1345	1467	1589	1711
3/8	357	407	458	509	611	713	815	917	1019	1120	1222	1324	1426
7/16	306	349	393	437	524	611	698	786	873	960	1048	1135	1222
1/2	267	306	344	382	458	535	611	688	764	840	917	993	1070

Figure 3-35. Partial listing of the relationship between cutting speed, spindle speed, and diameter in English units. (Butterfield)

Diameter in mm	CUTTING SPEED IN m/min													
	8	9	10	12	15	20	22	25	27	30	35	40	45	50
	Revolutions-per-minute													
5	510	573	636	764	955	1272	1398	1590	1720	1912	2230	2548	2870	3180
6	425	478	531	636	797	1060	1165	1325	1432	1593	1856	2124	2390	2650
7	364	409	455	516	683	910	1000	1136	1230	1365	1593	1820	2050	2275
8	318	358	400	478	597	796	876	996	1075	1191	1393	1592	1791	1990
9	288	318	354	425	530	708	780	886	955	1060	1240	1415	1590	1770
10	255	287	318	382	478	637	700	796	860	956	1125	1274	1435	1590
11	231	260	289	347	434	580	636	724	781	868	1013	1157	1300	1445
12	212	239	265	318	398	531	584	663	716	796	928	1060	1195	1325
13	196	220	245	294	367	490	539	612	662	735	857	980	1100	1225
14	182	205	228	273	341	455	502	568	615	682	796	910	1025	1136
15	169	191	212	254	313	425	467	531	572	635	740	846	952	1058
16	159	179	199	239	298	398	438	497	538	597	695	796	896	995
17	150	169	188	225	281	376	414	470	508	563	657	752	845	940
18	142	159	177	212	265	354	390	443	478	530	620	708	795	885
19	134	151	168	201	252	336	370	420	454	504	589	672	756	810
20	128	143	159	191	239	319	351	398	430	478	558	637	716	795
21	122	136	152	182	228	305	334	380	410	456	532	608	683	759
22	116	130	145	174	217	290	318	362	390	434	506	579	660	723
23	111	125	139	167	208	278	305	347	374	416	485	555	624	693
24	106	120	133	159	199	266	292	332	358	398	464	530	598	663
25	102	115	128	153	192	255	281	319	344	383	446	510	574	638

Figure 3-36. Partial listing of the relationship between cutting speed, spindle speed, and diameter in metric units.

Example 3-1: A drill 3/8" in diameter is used to drill a cast-iron plate. Determine the number of revolutions-per-minute at which the drill press spindle must be set for this job if the recommended cutting speed is 80 feet-per-minute.

Given: D = 3/8" (0.375"), CS = 80 ft/min

$$\text{Solution: } N = \frac{12 \times CS}{\pi \times D} = \frac{12 \times 80}{3.14 \times 0.375} = \frac{960}{1.1775} = 815 \text{ RPM}$$

Example 3-2: A drill 10 mm in diameter, is used to drill a piece of alloy steel. Determine the number of revolutions-per-minute at which the drill press spindle must be set for this job if the recommended cutting speed is 20 meters-per-minute.

Given: D = 10 mm, V = 20 m/min

$$\text{Solution: } N = \frac{1{,}000 \times V}{\pi \times D} = \frac{1{,}000 \times 20}{3.14 \times 10} = \frac{20{,}000}{31.4} = 637 \text{ U/min}$$

Machining Time

Machining time is the time required for the completion of a machining process, excluding the time involved in tool grinding and set-up. The machining time depends upon the spindle speed and feed. Machining time can be found by applying the following formula:

$$t = \frac{L}{N \times f}$$

t = time required to complete one pass in minutes

L = length or thickness of workpiece in inches or millimeters (mm)

N = spindle speed in revolutions-per-minute (RPM) or strokes-per-minute

f = feed in inches-per-revolution (ipr), or millimeters-per-revolution (mmpr), or inches-per-stroke (ips)

See Examples of how to estimate machining time under sections DRILL PRESS, LATHE, and MILLING MACHINE.

Cutting-tool Materials

The cutting tools used in machining processes are made of carbon steels, high-speed steels (HSS), cast alloys (STELLITE), cemented carbides, ceramics, and diamonds. The most commonly used tools, however, are those made of *high-speed steels* and *cemented carbides*. These two materials are characterized by hot hardness, toughness, and wear resistance, and are suitable for any machining process.

High-speed Steel Tool Materials. High-speed steels are special-alloy steels. The three types of high-speed steel tool materials are *tungsten high-speed steels, molybdenum high-speed steels,* and *cobalt high-speed steels.* Tungsten and molybdenum high-speed steels are used for general purpose and for abrasion resistant materials, while cobalt high-speed steels are used for heavy cuts and for abrasion-resistant hard materials.

Cemented-carbide Tool Materials. Cemented carbides are compounds of carbon and various non-ferrous metals such as titanium, vanadium, chromium, zirconium, molybdenum, tantalum, and tungsten. The simplest form of cemented carbide contains tungsten, carbon, and cobalt. Cemented-carbide tools are made from metal powders that are *sintered* (heated without melting), not melted or cast. They are capable of machining the hardest metals or alloys. In general, they are used in high-volume production, and in rigid and high-powered machine tools.

Cemented-carbide tools are available in a variety of shapes and grades. Each of these grades has a standard composition identified by industry code symbols (C-1, C-2, C-3, C-4, C-5, etc.). Figure 3-37 shows the code numbers used by various manufacturers to identify comparable, but not identical grade composition.

	APPLICATION	INDUSTRY CODE	ADAMAS	CARBOLOY	CARMET	FIRTH LEACH	FIRTH STERLING	KENNA-METAL	NEW-COMER	TALIDE	VALENITE	VASCOLOY RAMET	WESSON	WILLEY
Cast Iron, Non-Ferrous and Non-Metallic Materials	Roughing	C-1	B	44A	CA-3	FA-5	H	K1	NC-4	C-89	VC-1	RA 68 VR 59	GS	E8 E13
	General Purpose	C-2	A	883 860	CA-4	FA-6 FA-61	HA	K6	NC-3	C-91	VC-2	2A5 VR54	GI	E6
	Finishing	C-3	AA	905	CA-7	FA-7	HE	K8	NC-2	C-93	VC-3	2A7	GA	E5
	Precision Finishing	C-4	AAA	999	CA-8	FA-8	HF	K11	NC-2 NC-1	C-95	VC-4	2A7	GF	E3
Steel and Steel Alloys	Roughing	C-5	434	370	CA-51	FT-3	T-04	KM	NS-6	S-88	VC-5	EE VR 77	WS	945
	Roughing Alloy steel	C-50	434	370	CA-610	FT-41 FT-5	TXH	K21	NS-4 NS-65	S-88	VC-5	VR 77 VR 75	26	8A
	General Purpose	C-6	D	78B	CA-609	FT-4	TXH TA	K2S K21	NS-3	S-90	VC-6	VR 75	WM	710
	Finishing	C-7	C	78	CA-608	FT-6	T-16	K3H K4H	NS-2	S-92	VC-7	E VR 75	WH	606
	Semi-finishing & Finishing Alloy Steel	C-70	548	350	CA-606	FT-61	TXL	K4H K5H	NS-17	S-92	VC-7	VR 73	26	6A
	Precision Finishing	C-8	CC	330	CA-605	FT-7	T-31	K7H	NS-15 NM-95	S-94	VC-8	EH	WH	509

ROUGHING An extreme case of a roughing cut, which involves interrupted cuts, chilled surfaces or similar complicating factors.

GENERAL PURPOSE Initial machining operation involving the original surface of a bar, casting or forging. Removal of excess metal is the prime consideration.

SEMI-FINISHING The machining operation following a roughing cut or the initial cut on parts having clean surfaces and relatively close tolerances.

FINISHING The final machining operation to bring the part to blueprint requirements for size and finish.

PRECISION FINISHING An exact final machining operation to produce surfaces to close tolerances, together with a fine finish.

NOTE: THE ABOVE CHART IS NOT A GRADE COMPARISON CHART, NOR IS IT AN ENDORSEMENT OF ANY MANUFACTURER'S PRODUCT, OR AN APPROVED LIST OF SERVICES.

Figure 3-37. Identification of carbide cutting tool materials and their applications. (Cincinnati Milacron)

Cutting Fluids

Although metals can be machined dry, the use of a cutting fluid facilitates machining processes. Cutting fluids are oil-base liquids that reduce friction, dissipate heat, eliminate the development of excessive buildup at the tip of the cutting tool edge, and prevent metallic wear of moving parts. A variety of cutting fluids are available that may be used in machining operations. See Figure 3-38.

MACHINE TOOLS

Machine tools are power-driven tools used for various machining processes. Among the common machine tools used for general purpose machining operations are the *drill press, shaper and planer, lathe, boring machine, milling machine,* and *grinding machine.*

MATERIAL	DRILLING	REAMING	THREADING	TURNING	MILLING
Aluminum	Soluble oil Kerosene Kerosene and lard oil	Soluble oil Kerosene Mineral oil	Soluble oil Kerosene and lard oil	Soluble oil	Soluble oil Lard oil Mineral oil Dry
Brass	Dry Soluble oil Kerosene and lard oil	Dry Soluble oil	Soluble oil Lard oil	Soluble oil	Dry Soluble oil
Bronze	Dry Soluble oil Mineral oil Lard oil	Dry Soluble oil Mineral oil Lard oil	Soluble oil Lard oil	Soluble oil	Dry Soluble oil Mineral oil Lard oil
Cast iron	Dry Air jet Soluble oil	Dry Soluble oil Mineral lard oil	Dry Sulphurized oil Mineral lard oil	Dry Soluble oil	Dry Soluble oil
Copper	Dry Soluble oil Mineral lard oil Kerosene	Soluble oil Lard oil	Soluble oil Lard oil	Soluble oil	Dry Soluble oil
Malleable iron	Dry Soda water	Dry Soda water	Lard oil Soda water	Soluble oil	Dry Soda water
Monel metal	Soluble oil Lard oil	Soluble oil Lard oil	Lard oil	Soluble oil	Soluble oil
Steel alloys	Soluble oil Sulphurized oil Mineral lard oil	Soluble oil Sulphurized oil Mineral lard oil	Sulphurized oil Lard oil	Soluble oil	Soluble oil Mineral lard oil
Steel, machine	Soluble oil Sulphurized oil Lard oil Mineral lard oil	Soluble oil Mineral lard oil	Soluble oil Mineral lard oil	Soluble oil	Soluble oil Mineral lard oil
Steel, tool	Soluble oil Sulphurized oil Mineral lard oil	Soluble oil Sulphurized oil Lard oil	Sulphurized oil Lard oil	Soluble oil	Soluble oil Lard oil

Figure 3-38. Recommended cutting fluids for various materials and processes. (Cincinnati Milacron)

Drill Press

The drill press (Figure 3-39) is the basic machine tool used for drilling. In drilling operations, cutting takes place when the spindle with the cutting tool is rotating (cutting speed), while at the same time, the tool is fed slowly into the workpiece (feed). See Figure 3-40 for recommended cutting speed and feed for drilling operations.

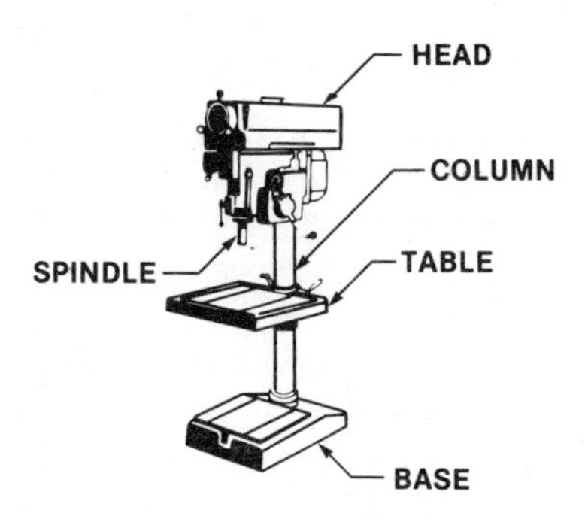

Figure 3-39. Drill press basic nomenclature. (Clausing Division of Atlas Press Company)

MATERIAL	CUTTING SPEED	
	f/min	m/min
Low-carbon steel	80-110	24-34
Medium-carbon steel	60-80	18-24
High-carbon steel	50-70	15-20
Alloy steel (soft)	50-60	15-18
Alloy steel (hard)	30-40	10-12
Cast iron (soft)	100-150	30-45
Cast iron (hard)	70-100	21-30
Aluminum alloys	200-300	60-90
Copper alloys	180-240	54-75

CUTTING FEED	
DIAMETER (in)	FEED (ipr)
up to 1/8	.001 - .002
1/8 - 1/4	.002 - .004
1/4 - 1	.004 - .007
1/2 - 1	.007 - .015

CUTTING FEED	
DIAMETER (mm)	FEED (mm/rev)
up to 5	0.11 - 0.16
5-12	0.16 - 0.32
12 - 25	0.32 - 0.50

Figure 3-40. Recommended cutting speed and feed for drilling.

Machining Time for Drilling. The machining time required for drilling a hole can be determined by using the following formula:

$$t = \frac{L}{N \times f}$$

See Example 3-3 for an application of this formula.

Example 3-3: A drill 5/16″ in diameter is used to drill a 2½″ thick hard alloy steel plate. Determine the time required if the recommended cutting speed is 30 feet-per-minute and the feed is 0.004 inch-per-revolution.

Given: D = 5/16″ (0.3125″), CS = 30 ft/min, L = 2½″ (2.5″), f = 0.004 ipr

Solution: $$N = \frac{12 \times CS}{\pi \times D} = \frac{12 \times 30}{3.14 \times 0.3125} = \frac{360}{0.98125} = 367 \text{ RPM}$$

$$t = \frac{L}{N \times f} = \frac{2.5}{367 \times 0.004} = \frac{2.5}{1.468} = 1.7 \text{ min}$$

Drill Press Cutting Tools. A drill press uses multi-edge cutting tools (Figure 3-41). Of all the cutting tools used in drill press work, only the two-edge (two-lip) drill is used to make holes. The other multi-edge cutting tools are used to further machine existing holes.

Twist Drill. The twist drill, or drill (Figure 3-42), is one of the most important cutting tools used in machine shop work. It is used in the drill press, milling machine, lathe, and portable electric drill for making holes in metals and other materials.

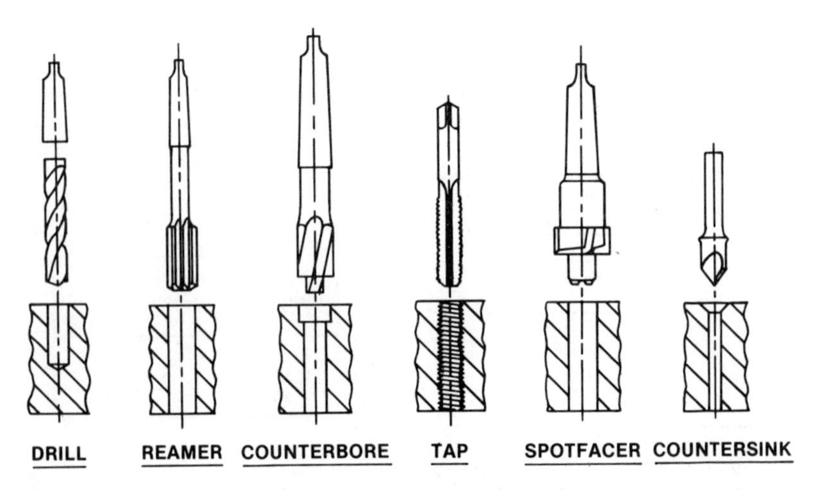

Figure 3-41. Drill press cutting tools.

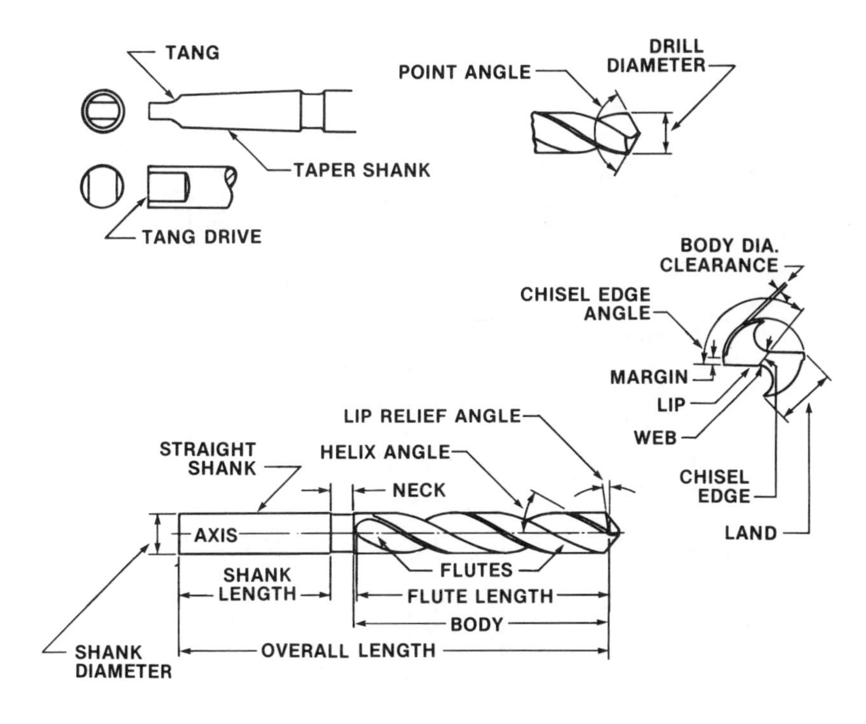

Figure 3-42. Twist drill nomenclature. (Standard Tool Div. LSI)

Drill Sizes. Drill sizes have been standardized and are available in *number sizes, letter sizes, fraction sizes,* and *metric sizes.* The common sizes available in these catergories are:

1. Number sizes: 80 to 1, ranging in sizes from .0135″ to .2280″.
2. Letter sizes: A to Z, ranging in sizes from .234″ to .413″.
3. Fraction sizes: 1/64″ to 1 3/4″ in 1/64″ increments; 1 3/4″ to 2 1/4″ in 1/32″ increments; and 2 1/4″ to 3 1/2″ in 1/16″ increments.
4. Metric sizes: 0.15 mm to 60.00 mm.

See Figure 3-43 for a typical inch drill-size chart, and Figure 3-44 for a typical metric drill-size chart with inch equivalents.

Holes 3/8″ or larger must be drilled with a smaller drill first, then with the 3/8″ or larger size drill. The diameter of the smaller size drill (pilot drill) should be approximately equal to the larger drill's web dimension. For effective drilling, the point of the drill must be ground properly to suit the material. Figure 3-45 shows the suggested drill point specifications for drilling various materials.

DRILL NO.	FRAC.	DECI.	DRILL NO.	FRAC.	DECI.	DRILL NO.	FRAC.	DECI.	DRILL NO.	FRAC.	DECI.
80	—	.0135	42	—	.0935	7	—	.201	X	—	.397
79	—	.0145	—	3/32	.0938	—	13/64	.203	Y	—	.404
—	7/64	.0156				6	—	.204			
78	—	.0160	41	—	.0960	5	—	.206	—	13/32	.406
77	—	.0180	40	—	.0980	4	—	.209	Z	—	.413
			39	—	.0995				—	27/64	.422
76	—	.0200	38	—	.1015	3	—	.213	—	7/16	.438
75	—	.0210	37	—	.1040	—	7/32	.219	—	29/64	.453
74	—	.0225				2	—	.221			
73	—	.0240	36	—	.1065	1	—	.228	—	15/32	.469
72	—	.0250	—	7/64	.1094	A	—	.234	—	31/64	.484
			35	—	.1100				—	1/2	.500
71	—	.0260	34	—	.1110	—	15/64	.234	—	33/64	.516
70	—	.0280	33	—	.1130	B	—	.238	—	17/32	.531
69	—	.0292				C	—	.242			
68	—	.0310	32	—	.116	D	—	.246	—	35/64	.547
—	1/32	.0313	31	—	,120	—	1/4	.250	—	9/16	.562
			—	1/8	.125				—	37/64	.578
67	—	.0320	30	—	,129	E	—	.250	—	19/32	.594
66	—	.0330	29	—	.136	F	—	.257	—	39/64	.609
65	—	.0350				G	—	.261			
64	—	.0360	—	9/64	.140	—	17/64	.266	—	5/8	.625
63	—	.0370	28	—	.141	H	—	.266	—	41/64	.641
			27	—	.144				—	21/32	.656
62	—	.0380	26	—	.147	I	—	.272	—	43/64	.672
61	—	.0390	25	—	.150	J	—	.277	—	11/16	.688
60	—	.0400				—	9/32	.281			
59	—	.0410	24	—	.152	K	—	.281	—	45/64	.703
58	—	.0420	23	—	.154	L	—	.290	—	23/32	.719
			—	5/32	.156				—	47/64	.734
57	—	.0430	22	—	.157	M	—	.295	—	3/4	.750
56	—	.0465	21	—	.159	—	19/64	.297	—	49/64	.766
—	3/64	.0469				N	—	.302			
55	—	.0520				—	5/16	.313	—	25/32	.781
54	—	.0550	20	—	.161	O	—	.316	—	51/64	.797
			19	—	.166				—	13/16	.813
53	—	.0595	18	—	.170	P	—	.323	—	53/64	.828
—	1/16	.0625	—	11/64	.172	—	21/64	.328	—	27/32	.844
52	—	.0635	17	—	.173	Q	—	.332			
51	—	.0670				R	—	.339			
50	—	.0700	16	—	.177	—	11/32	.344	—	55/64	.859
			15	—	.180				—	7/8	.875
49	—	.0730	14	—	.182	S	—	.348	—	57/64	.891
48	—	.0760	13	—	.185	T	—	.358	—	29/32	.905
—	5/64	.0781	—	3/16	.188	—	23/64	.359	—	59/64	.922
47	—	.0785				U	—	.368			
46	—	.0810	12	—	.189	—	3/8	.375	—	15/16	.938
			11	—	.191				—	61/64	.953
45	—	.0820	10	—	.194	V	—	.377	—	31/32	.969
44	—	.0860	9	—	.196	W	—	.386	—	63/64	.984
43	—	.0890	8	—	.199	—	25/64	.391	—	1	1.000

Figure 3-43. American National standard drill sizes. (The L. S. Starrett Company)

DRILL DIAMETER		DRILL DIAMETER		DRILL DIAMETER		DRILL DIAMETER	
mm	in.	mm	in.	mm	in.	mm	in.
.15	.0059	.69	.0272	3.25	.1280	7.75	.3051
.17	.0067	.71	.0280	3.40	.1339	7.90	.3110
.19	.0075	.73	.0287	3.60	.1417	8.10	.3189
.21	.0083	.75	.0295	3.75	.1476	8.25	.3243
.23	.0091	.77	.0303	3.90	.1535	8.40	.3307
.25	.0098	.79	.0311	4.10	.1614	8.60	.3386
.27	.0106	.85	.0335	4.25	.1673	8.75	.3445
.29	.0114	.95	.0374	4.40	.1732	8.90	.3504
.31	.0122	1.05	.0413	4.60	.1811	9.10	.3583
.33	.0130	1.15	.0453	4.75	.1870	9.25	.3642
.35	.0139	1.25	.0492	4.90	.1929	9.40	.3701
.37	.0146	1.35	.0532	5.10	.2008	9.70	.3819
.39	.0154	1.45	.0571	5.25	.2067	9.80	.3898
.41	.0161	1.55	.0610	5.40	.2126	10.50	.4134
.43	.0169	1.65	.0650	5.60	.2205	11.50	.4528
.45	.0177	1.75	.0689	5.75	.2264	12.50	.4921
.47	.0185	1.85	.0728	5.90	.2323	13.50	.5315
.49	.0193	1.95	.0768	6.10	.2402	14.50	.5709
.51	.0201	2.05	.0807	6.25	.2461	15.50	.6102
.53	.0209	2.15	.0846	6.40	.2520	16.50	.6496
.55	.0217	2.25	.0886	6.60	.2598	17.50	.6890
.57	.0224	2.35	.0906	6.75	.2657	18.50	.7283
.59	.0232	2.45	.0965	6.90	.2717	19.50	.7677
.61	.0240	2.60	.1024	7.10	.2795	20.50	.8071
.63	.0248	2.75	.1083	7.25	.2854	21.50	.8465
.65	.0256	2.90	.1142	7.40	.2913	23.00	.9055
.67	.0264	3.10	.1220	7.60	.2992	25.00	.9843

Figure 3-44. Partial listing of metric standard drill sizes with decimal-inch equivalents.

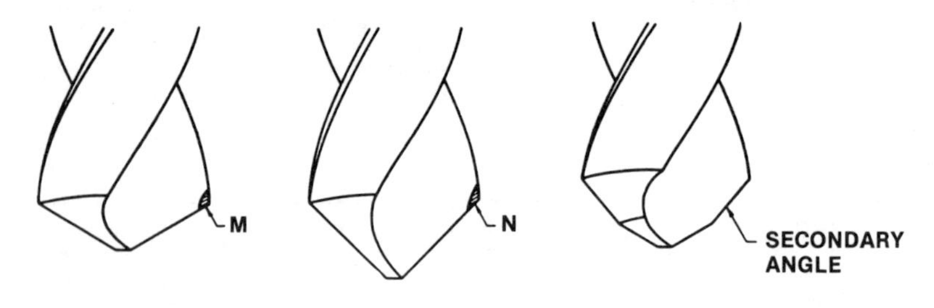

DRILL POINTS FOR CAST IRON

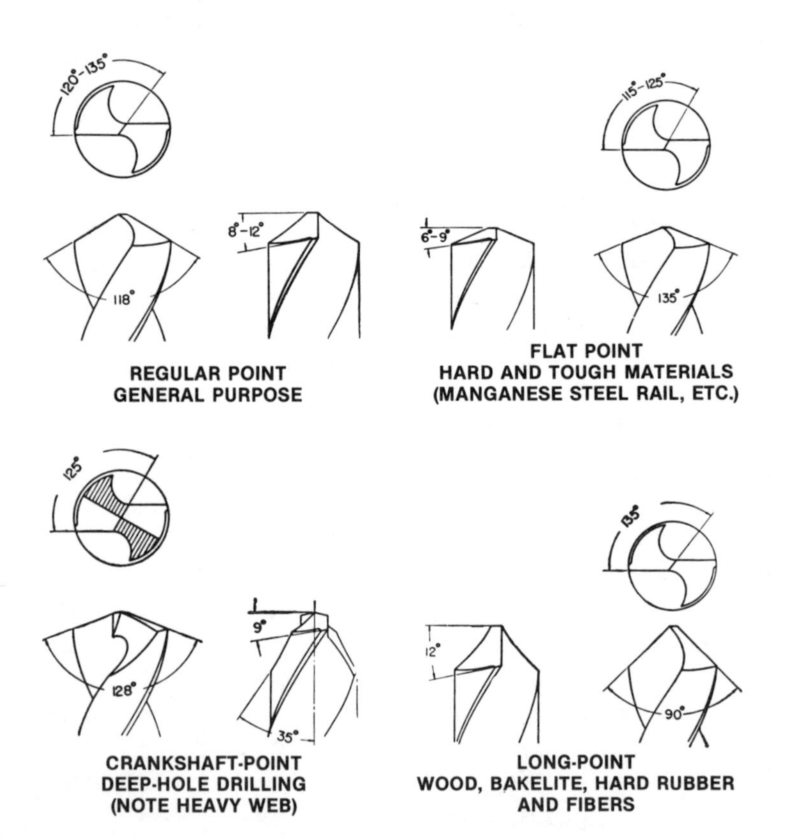

NOTE: WHEN DRILLING CAST IRON, IF A DRILL IS SHARPENED WITH A LARGE INCLUDED POINT ANGLE, THE ZONE OF ABRASION IS COMPARATIVELY SMALL, AS IN M. THIS RESULTS IN FASTER DRILLING, BUT SHORT TOOL LIFE. BY SHARPENING THE DRILL WITH A SMALL INCLUDED POINT ANGLE, THE ZONE OF ABRASION N IS INCREASED AND WITH IT, THE LIFE OF THE DRILL IS INCREASED. TO FURTHER INCREASE THE ABRASIVE AREA AND THE TOOL LIFE, IT IS SOMETIMES ADVISABLE TO GRIND A SECONDARY ANGLE AT THE CORNERS OF THE DRILL.

Figure 3-45. Suggested drill point specifications for drilling various materials. (Cleveland Twist Drill Company)

Any difficulty encountered in drilling has a specific cause. The following checklist shows common drilling problems and their causes:

CHECKLIST OF COMMON DRILL PROBLEMS AND CAUSES

PROBLEMS	CAUSES
Outer corners, breakdown	Cutting speed too high. Hard spots in material. No cutting compound at drill point. Flutes clogged with chips.
Cutting edges chip	Too much feed. Lip clearance too great.
Checks or cracks in cutting edges	Overheated or too quickly cooled while sharpening or drilling.
Margin chips	Oversize jig bushing.
Drill breaks	Point improperly ground. Feed too heavy. Spring or backlash in drill press, fixture, or work. Drill is dull. Flutes clogged with chips.
Tang breaks	Imperfect fit between taper shank and socket caused by dirt or chips, or burred or badly worn sockets.
Drill breaks when drilling brass or wood	Flutes clogged with chips. Improper type drill.
Drill splits up center	Lip clearance too small. Too much feed.
Drill will not enter work	Drill is dull. Lip clearance too small. Too heavy a web.
Hole rough	Point improperly ground or dull. No cutting compound at drill point. Improper cutting compound. Feed too great. Fixture not rigid.
Hole oversize	Unequal angle or length of the cutting edges—or both. Loose spindle.
Chip shape changes while drilling	Drill becomes dull or cutting edges chipped.
Large chip coming out of one flute, small chip out of other flute	Point improperly ground, one lip doing all the cutting.

(Cleveland Twist Drill Company)

Drills and Taps. One of the most important uses of drills is to make holes for tapping, or cutting internal threads. The size of the tap drill depends on the diameter of the fastener and the type of thread. The size of the clearance drill depends only on the diameter of the screw (nominal size).

Standard Screw Threads for Fasteners

An *external thread* is the helical, or spiral ridge on a screw. An *internal thread* is the helical, or spiral ridge in the nut. Several forms of threads exist (Figure 3-46), each of which has unique characteristics. Basic thread terminology is shown in Figure 3-47. Among the most common threads used for fasteners are the *Unified National, American National,* and *International Organization for Standardization* (ISO) *Metric* threads.

Unified National and American National Threads. The Unified National thread was adopted in 1946 by the United States, Great Britain, and Canada. It is essentially the same as the American National thread. The distinguishing characteristic is that unlike the American National thread, the root of the Unified National thread is rounded instead of flat. Unified National and American National threads are interchangeable. Both the Unified National and the American National threads are available in four series of major diameter-pitch combinations. They are distinguished from each other by their pitch. The four series are:

1. NC or UNC: National Coarse or Unified National Coarse
2. NF or UNF: National Fine or Unified National Fine
3. NS or UNS: National Special or Unified National Special
4. NPT: Taper Pipe Thread

Thread Classes. Unified National and American National threads are available in three classes. The classes of Unified National threads are designated by numbers and letters (1A, 2A, or 3A for external threads, and 1B, 2B, or 3B for internal threads). The classes of American National threads are designated by numbers (1, 2, or 3 for either external or internal threads). Classes 1, 1A, and 1B threads have little tolerance and allowance and are used for precision work. Most of the commercial fasteners are Classes 2, 2A, 2B, 3, 3A, and 3B.

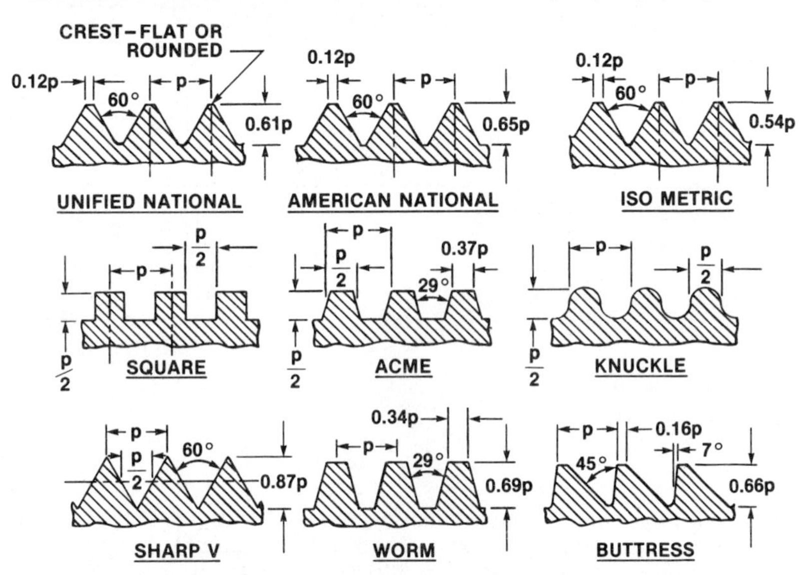

Figure 3-46. Standard forms of screw threads.

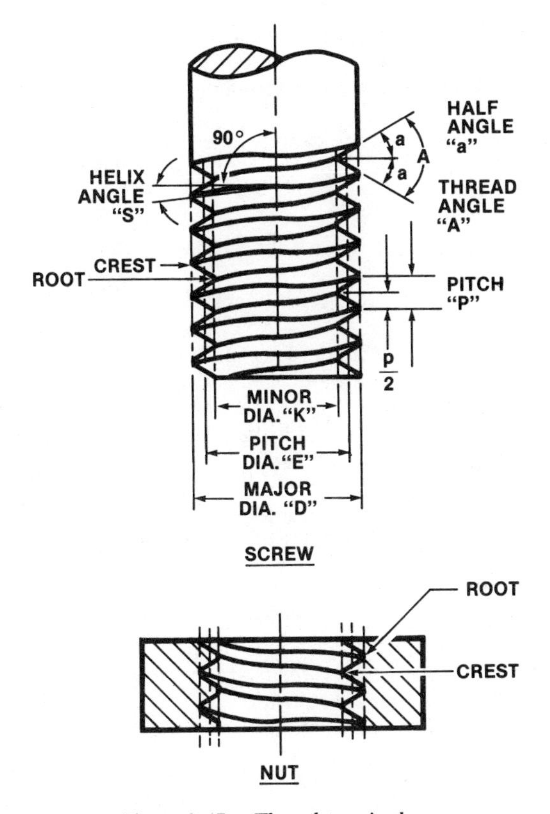

Figure 3-47. Thread terminology.

Thread Designation. Unified National and American National thread specifications are designated in the following sequence:

1. Nominal size: major or largest diameter of either internal or external thread in inches.
2. Number of threads-per-inch (n). NOTE: $1/n''$ = pitch (p)
3. Thread series symbol: UNC, UNF, and UNS for Unified National threads; and NC, NF, NS, and NPT for American National threads.
4. Thread class: Class 1, 2, or 3 for American National threads; and Class 1A, 1B, 2A, 2B, 3A, and 3B for Unified National threads.

The following are examples of thread specifications for Unified National and American National threads:

UNIFIED NATIONAL	AMERICAN NATIONAL
3/8 - 16 UNC - 2B	3/8 - 16 NC - 2
3/8 - nominal size	3/8 - nominal size
16 - threads-per-inch	16 - threads-per-inch
UNC - Unified National Coarse	NC - National Coarse
2B - Class 2 (internal thread)	2 - Class 2, either external or internal

ISO Metric Threads. The ISO metric screw threads, like the Unified National and American National threads, have a 60° angle thread. Their dimensions, however, are in millimeters and are designated in the following sequence:

1. Nominal size preceded by the capital letter "M".
2. Pitch of the thread preceded by the capital letter "X".
3. Thread class symbol.

The following is an example of an ISO metric thread specification:

M20X2-5g6g

1. M20: nominal size or major diameter is 20 mm.
2. X2: pitch of the thread is 2 mm.
3. 5g6g: thread class symbol - 5g is the tolerance grade and tolerance position specified for the pitch diameter; 6g is the tolerance grade and tolerance position specified for the major diameter.

NOTE: For additional information regarding ISO tolerance grade and tolerance position, see STANDARD ISO FITS in this chapter.

Three classes of ISO metric threads designated by tolerance grades 4, 6, and 8 exist. There are also tolerance positions e, g, and h for external threads, and tolerance positions G and H for internal threads.

The most common ISO metric threads are made to specifications provided by tolerance grade 6 and tolerance position g. This combination corresponds to the Unified National thread Class 2A and 2B.

Tap and Clearance Drill Sizes. A *tap drill* is used to make holes for tapping. A *clearance drill* is used to make holes large enough to allow bolts to pass through the holes freely. Therefore, the clearance drill size should be slightly larger than the nominal size (major diameter) of the bolt, and the tap drill size should be smaller than the major diameter of the bolt, but slightly larger than its minor diameter. Figure 3-48 shows the relationship between clearance drill diameter, minor diameter, and tap drill diameter for 75% full thread.

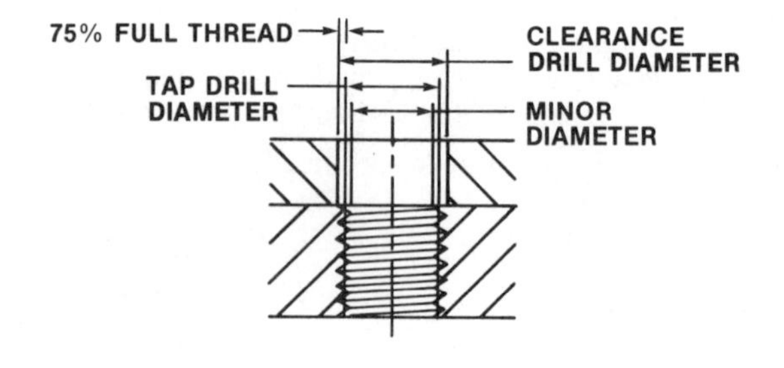

Figure 3-48. Relative size of tap drill diameter for 75% full thread.

The use of a tap drill (larger than the minor diameter screw thread size) will produce a thread that is less than 100% full. This, however, does not affect the strength of the thread if it is kept within limits. The difference between minor diameter and tap drill diameter provides clearance for the chips that are created in tapping, and allows for the free turning of the tap. Most interchangeable screw threads have approximately 75% full thread. See Figures 3-49 and 3-50 for tap drill sizes and other information related to Unified National and American National threads, and ISO metric threads, respectively.

Lathe

The lathe (Figure 3-51) is the basic machine tool used for turning. In turning operations, cutting takes place when the *spindle* with the workpiece is rotating. The carriage provides the feed by moving the cutting tool along or across the *bed*. Figure 3-52 shows the recommended cutting speed and cutting tool angles for turning and boring. Figure 3-53 shows recommended feed and depth-of-cut.

Machining Time for Turning and Boring. The time required for any machining operation in lathe work depends upon the spindle speed (RPM), the feed (f) in ipr or mmpr, and the length of the workpiece (L). See Example 3-4.

Example 3-4: A shaft 18″ long and 2″ in diameter is to be machined in a general purpose lathe. Determine the time required for one pass if the recommended speed is 60 feet-per-minute and the recommended feed is 0.010 inch-per-revolution.

Given: $L = 18''$, $D = 2''$, $CS = 60$ ft/min, $f = 0.010$ ipr

Solution: $$N = \frac{12 \times CS}{\pi \times D} = \frac{12 \times 60}{3.14 \times 2} = \frac{720}{6.28} = 115 \text{ RPM}$$

$$t = \frac{L}{N \times f} = \frac{18}{115 \times 0.01} = \frac{18}{1.15} = 15.6 \text{ min}$$

Lathe Cutting Tools. Lathe cutting tools are made from various cutting-tool materials and are available in a variety of shapes and sizes. Among the frequently used lathe cutting tools are those made from *high-speed steel (HSS)* and *cemented carbides*.

High-speed Steel Lathe Cutting Tools. Lathe cutting tools (tool bits) made from high-speed steel are available in sizes ranging from 3/16″ × 3/16″ × 2″ to 5/8″ × 5/8″ × 4 1/2″. For effective machining, these tools must be ground properly to form certain angles (Figure 3-54).

The machining operation involved determines the shape of a tool bit to be used. See Figure 3-55 for the application of commonly used tool bit shapes.

Cemented-carbide Lathe Cutting Tools. Lathe cutting tools made from carbides are smaller than HSS cutting tools. They are available in standard sizes and shapes and are either brazed or indexable (Figure 3-56).

For effective machining with cemented-carbide tools, it is imperative to follow the manufacturer's recommendations for use.

TAP	TAP DRILL	DECIMAL EQUIV. OF TAP DRILL	THEORETICAL % OF THREAD	PROBABLE OVERSIZE (MEAN)	PROBABLE HOLE SIZE	PERCENTAGE OF THREAD	TAP	TAP DRILL	DECIMAL EQUIV. OF TAP DRILL	THEORETICAL % OF THREAD	PROBABLE OVERSIZE (MEAN)	PROBABLE HOLE SIZE	PERCENTAGE OF THREAD
0 - 80	56	.0465	83	.0015	.0480	74		14	.1820	73	.0035	.1855	66
	3/64	.0469	81	.0015	.0484	71		13	.1850	67	.0035	.1885	59
1 - 64	54	.0550	89	.0015	.0565	81		3/16	.1875	61	.0035	.1910	54
	53	.0595	67	.0015	.0610	59	1/4 - 20	9	.1960	83	.0038	.1998	77
1 - 72	53	.0595	75	.0015	.0610	67		8	.1990	79	.0038	.2028	73
	1/16	.0625	58	.0015	.0640	50		7	.2010	75	.0038	.2048	70
2 - 56	51	.0670	82	.0017	.0687	74		13/64	.2031	72	.0038	.2069	66
	50	.0700	69	.0017	.0717	62		6	.2040	71	.0038	.2078	65
	49	.0730	56	.0017	.0747	49		5	.2055	69	.0038	.2093	63
2 - 64	50	.0700	79	.0017	.0717	70		4	.2090	63	.0038	.2128	57
	49	.0730	64	.0017	.0747	56	1/4 - 28	3	.2130	80	.0038	.2168	72
3 - 48	48	.0760	85	.0019	.0779	78		7/32	.2188	67	.0038	.2226	59
	5/64	.0781	77	.0019	.0800	70		2	.2210	63	.0038	.2248	55
	47	.0785	76	.0019	.0804	69	5/16 - 18	F	.2570	77	.0038	.2608	72
	46	.0810	67	.0019	.0829	60		G	.2610	71	.0041	.2651	66
	45	.0820	63	.0019	.0839	56		17/64	.2656	65	.0041	.2697	59
3 - 56	46	.0810	78	.0019	.0829	69		H	.2660	64	.0041	.2701	59
	45	.0820	73	.0019	.0839	65	5/16 - 24	H	.2660	86	.0041	.2701	78
	44	.0860	56	.0019	.0879	48		I	.2720	75	.0041	.2761	67
4 - 40	44	.0860	80	.0020	.0880	74		J	.2770	66	.0041	.2811	58
	43	.0890	71	.0020	.0910	65	3/8 - 16	5/16	.3125	77	.0044	.3169	72
	42	.0935	57	.0020	.0955	51		O	.3160	73	.0044	.3204	68
	3/32	.0938	56	.0020	.0958	50		P	.3230	64	.0044	.3274	59
4 - 48	42	.0935	68	.0020	.0955	61	3/8 - 24	21/64	.3281	87	.0044	.3325	79
	3/32	.0938	68	.0020	.0958	60		Q	.3320	79	.0044	.3364	71
	41	.0960	59	.0020	.0980	52		R	.3390	67	.0044	.3434	58
5 - 40	40	.0980	83	.0023	.1003	76	7/16 - 14	T	.3580	86	.0046	.3626	81
	39	.0995	79	.0023	.1018	71		23/64	.3594	84	.0046	.3640	79
	38	.1015	72	.0023	.1038	65		U	.3680	75	.0046	.3726	70
	37	.1040	65	.0023	.1063	58		3/8	.3750	67	.0046	.3796	62
5 - 44	38	.1015	79	.0023	.1038	72		V	.3770	65	.0046	.3816	60
	37	.1040	71	.0023	.1063	63	7/16 - 20	W	.3860	79	.0046	.3906	72
	36	.1065	63	.0023	.1088	55		25/64	.3906	72	.0046	.3952	65

continued

TAP	TAP DRILL	DECIMAL EQUIV. OF TAP DRILL	THEORETICAL % OF THREAD	PROBABLE OVERSIZE (MEAN)	PROBABLE HOLE SIZE	PERCENTAGE OF THREAD	TAP	TAP DRILL	DECIMAL EQUIV. OF TAP DRILL	THEORETICAL % OF THREAD	PROBABLE OVERSIZE (MEAN)	PROBABLE HOLE SIZE	PERCENTAGE OF THREAD
6 - 32	37	.1040	84	.0023	.1063	78		X	.3970	62	.0046	.4016	55
	36	.1065	78	.0026	.1091	71	1/2 - 13	27/64	.4219	78	.0047	.4266	73
	7/64	.1094	70	.0026	.1120	64		7/16	.4375	63	.0047	.4422	58
	35	.1100	69	.0026	.1126	63	1/2 - 20	29/64	.4531	72	.0047	.4578	65
	34	.1110	67	.0026	.1136	60	9/16 - 12	15/32	.4688	87	.0048	.4736	82
	33	.1130	62	.0026	.1156	55		31/64	.4844	72	.0048	.4892	68
6 - 40	34	.1110	83	.0026	.1136	75	9/16 - 18	1/2	.5000	87	.0048	.5048	80
	33	.1130	77	.0026	.1156	69		33/64	.5156	65	.0048	.5204	58
	32	.1160	68	.0026	.1186	60	5/8 - 11	17/32	.5313	79	.0049	.5362	75
8 - 32	29	.1360	69	.0029	.1389	62		35/64	.5469	66	.0049	.5518	62
	28	.1405	58	.0029	.1434	51	5/8 - 18	9/16	.5625	87	.0049	.5674	80
8 - 36	29	.1360	78	.0029	.1389	70		37/64	.5781	65	.0049	.5831	58
	28	.1405	68	.0029	.1434	57	3/4 - 10	41/64	.6406	84	.0050	.6456	80
	9/64	.1406	68	.0029	.1435	57		21/32	.6563	72	.0050	.6613	68
10 - 24	27	.1440	85	.0032	.1472	79	3/4 - 16	11/16	.6875	77	.0050	.6925	71
	26	.1470	79	.0032	.1502	74	7/8 - 9	49/64	.7656	76	.0052	.7708	72
	25	.1495	75	.0032	.1527	69		25/32	.7812	65	.0052	.7864	61
	24	.1520	70	.0032	.1552	64	7/8 - 14	51/64	.7969	84	.0052	.8021	79
	23	.1540	67	.0032	.1572	61		13/16	.8125	67	.0052	.8177	62
	5/32	.1563	62	.0032	.1595	56	1″ - 8	55/64	.8594	87	.0059	.8653	83
	22	.1570	61	.0032	.1602	55		7/8	.8750	77	.0059	.8809	73
10 - 32	5/32	.1563	83	.0032	.1595	75		57/64	.8906	67	.0059	.8965	64
	22	.1570	81	.0032	.1602	73		29/32	.9063	58	.0059	.9122	54
	21	.1590	76	.0032	.1622	68	1″ - 12	29/32	.9063	87	.0060	.9123	81
	20	.1610	71	.0032	.1642	64		59/64	.9219	72	.0060	.9279	67
	19	.1660	59	.0032	.1692	51		15/16	.9375	58	.0060	.9435	52
12 - 24	11/64	.1719	82	.0035	.1754	75	1″ - 14	59/64	.9219	84	.0060	.9279	78
	17	.1730	79	.0035	.1765	73		15/16	.9375	67	.0060	.9435	61
	16	.1770	72	.0035	.1805	66	1 1/8 - 7	31/32	.9688	84	.0062	.9750	81
	15	.1800	67	.0035	.1835	60		63/64	.9844	76	.0067	.9911	72
	14	.1820	63	.0035	.1855	56		1″	1.000	67	.0070	1.0070	64
12 - 28	16	.1770	84	.0035	.1805	77		1 1/64	1.0156	59	.0070	1.0226	55
	15	.1800	78	.0035	.1835	70							

Figure 3-49. Partial listing of top drill sizes for producing various percentages of thread for Unified National and American National threads. (Standard Tool Div. LSI)

METRIC TAP SIZE & PITCH	BASED ON APPROX. 60% THREAD				BASED ON APROX. 75% THREAD			
	Theoretical drill size mm	Recommended metric drill size/mm	Closest American drill		Theoretical drill size mm	Recommended metric drill size/mm	Closest American drill	
			size	mm equiv.			size	mm equiv.
M1.5 x 0.35	1.2272	1.20	3/64	1.1913	1.1590	1.15	57	1.0922
M1.6 x 0.35	1.3272	1.30	55	1.3208	1.2590	1.25	3/64	1.1913
M1.8 x 0.35	1.5272	1.50	53	1.5113	1.4590	1.45	54	1.3970
M2 x 0.45	1.6492	1.65	52	1.6129	1.5612	1.55	53	1.5113
M2 x 0.4	1.6882	1.65	52	1.6129	1.6102	1.60	1/16	1.5875
M2.2 x 0.45	1.8492	1.85	49	1.8542	1.7615	1.75	51	1.7018
M2.3 x 0.4	1.9882	1.95	5/64	1.9837	1.9102	1.90	49	1.8542
M2.5 x 0.45	2.1492	2.15	45	2.0828	2.0615	2.05	46	2.0574
M2.6 x 0.45	2.2492	2.25	44	2.1844	2.1615	2.15	45	2.0828
M3 x 0.6	2.5324	2.50	39	2.5273	2.4155	2.40	3/32	2.3825
M3 x 0.5	2.6147	2.60	38	2.5781	2.5184	2.50	40	2.4892
M3.5 x 0.6	3.0324	3.00	32	2.9464	2.9155	2.90	33	2.8702
M4 x 0.75	3.4154	3.40	30	3.2639	3.2693	3.25	30	3.2639
M4 x 0.7	3.4544	3.40	29	3.4544	3.3180	3.30	30	3.2639
M4.5 x 0.75	3.9154	3.90	23	3.9116	3.7693	3.75	26	3.7338
M5 x 1	4.2206	4.20	19	4.2164	4.0257	4.00	22	3.9878
M5 x 0.9	4.2988	4.25	19	4.2164	4.1235	4.10	20	4.0894
M5 x 0.8	4.3764	4.30	11/64	4.3662	4.2204	4.20	19	4.2164
M5.5 x 0.9	4.7988	4.75	3/16	4.7625	4.6235	4.60	14	4.6228
M6 x 1	5.2206	5.20	5	5.2197	5.0257	5.00	9	4.9784
M6 x 0.75	5.4154	5.40	3	5.4102	5.2693	5.25	5	5.2197
M7 x 1	6.2206	6.20	C	6.1468	6.0257	6.00	15/64	5.9537
M7 x 0.75	6.4154	6.40	D	6.2484	6.2693	6.25	D	6.2484
M8 x 1.25	7.0258	7.00	I	6.9088	6.7823	6.75	H	6.7564
M8 x 1	7.2206	7.20	9/32	7.1450	7.0257	7.00	I	6.9088
M9 x 1.25	8.0258	8.00	5/16	7.9375	7.7823	7.75	N	7.6708
M9 x 1	8.2206	8.20	P	8.2042	8.0257	8.00	5/16	7.9375
M10 x 1.5	8.8308	8.80	11/32	8.7325	8.5385	8.50	Q	8.4328
M10 x 1.25	9.0258	9.00	S	8.8392	8.7823	8.75	11/32	8.7325
M10 x 1	9.2206	9.20	23/64	9.1286	9.0257	9.00	S	8.8392
M11 x 1.5	9.8308	9.80	W	9.8044	9.5385	9.50	3/8	9.5250
M12 x 1.75	10.6360	10.50	Z	10.4902	10.2950	10.00	Y	10.2616
M12 x 1.5	10.8308	10.80	27/64	10.7162	10.5385	10.50	Z	10.4902
M12 x 1.25	11.0258	11.00	27/64	10.7162	10.7823	10.50	27/64	10.7162
M14 x 2	12.4413	12.40	31/64	12.3037	12.0516	12.00	15/32	11.9075
M14 x 1.5	12.8308	12.80	1/2	12.7000	12.5385	12.50	31/64	12.3037
M14 x 1.25	13.0258	13.00	1/2	12.7000	12.7823	12.70	1/2	12.7000
M15 x 1.5	13.8308	13.80	17/32	13.4950	13.5385	13.50	17/32	13.4950
M16 x 2	14.4413	14.25	9/16	14.2875	14.0516	14.00	35/64	13.8912
M16 x 1.5	14.8308	14.75	37/64	14.6837	14.5385	14.50	9/16	14.2875
M17 x 1.5	15.8308	15.75	39/64	15.4787	15.5385	15.50	39/64	15.4787
M18 x 2.5	16.0513	16.00	5/8	15.8750	15.5643	15.50	39/64	15.4787
M18 x 2	16.4413	16.25	41/64	16.2712	16.0516	16.00	5/8	15.8750
M18 x 1.5	16.8308	16.75	21/32	16.6700	16.5385	16.50	41/64	16.2712
M19 x 2.5	17.0513	17.00	21/32	16.6700	16.5643	16.50	41/64	16.2712
M20 x 2.5	18.0513	18.00	45/64	17.8587	17.5643	17.50	11/16	17.4625
M20 x 2	18.4413	18.25	23/32	18.2557	18.0516	18.00	45/64	17.8587
M20 x 1.5	18.8308	18.75	47/64	18.6538	18.5385	18.50	23/32	18.2557
M22 x 2.5	20.0513	20.00	25/32	19.8425	19.5643	19.50	49/64	19.4462
M22 x 2	20.4413	20.25	51/64	20.2413	20.0516	20.00	25/32	19.8425
M22 x 1.5	20.8308	20.75	13/16	20.6358	20.5385	20.50	51/64	20.2413
M24 x 3	21.6619	21.50	27/32	21.4325	21.0773	21.00	53/64	21.0337
M24 x 2	22.4413	22.25	7/8	22.2250	22.0516	22.00	55/64	21.8288
M24 x 1.5	22.8308	22.75	57/64	22.6212	22.5385	22.50	7/8	22.2250

APPROXIMATE METRIC TAP DRILL FORMULA

Nominal O.D. Minus (*Pitch*) = 77% of Thread

Nominal O.D. Minus (*.65 x Pitch*) = 50% of Thread	Nominal O.D. Minus (*.91 x Pitch*) = 70% of Thread
Nominal O.D. Minus (*.71 x Pitch*) = 55% of Thread	Nominal O.D. Minus (*.97 x Pitch*) = 75% of Thread
Nominal O.D. Minus (*.78 x Pitch*) = 60% of Thread	Nominal O.D. Minus (*1.04 x Pitch*) = 80% of Thread
Nominal O.D. Minus (*.84 x Pitch*) = 65% of Thread	Nominal O.D. Minus (*1.10 x Pitch*) = 85% of Thread

Figure 3-50. Partial listing of tap drill sizes for producing 60% and 75% ISO metric threads. (Regal-Beloit Corporation)

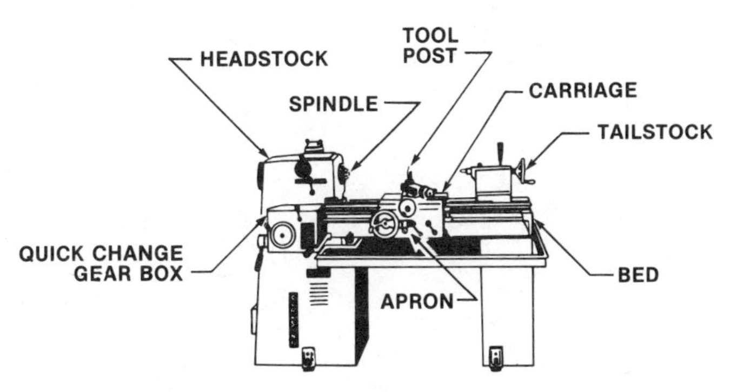

Figure 3-51. Basic lathe nomenclature. (Clausing Division of Atlas Press Company)

MATERIAL	CUTTING SPEED		ANGLES IN DEGREES			
FERROUS METALS	**ft/min**	**m/min**	**FC**	**SC**	**BR**	**SR**
Low-carbon steel	100 - 120	30 - 36	8	8	16½	22
Medium-carbon steel	80 - 100	24 - 30	8	8	12	14
High-carbon steel	60 - 80	18 - 24	8	8	8	12
Alloy steel (soft)	60 - 100	18 - 30	8	8	8	14
Alloy steel (hard)	30 - 60	9 - 18	8	8	8	12
Stainless steel	30 - 50	9 - 15	8	10	16½	10
Cast iron (soft)	80 - 100	24 - 36	8	8	8	14
Cast iron (hard)	40 - 60	12 - 18	8	8	5	10
NON-FERROUS METALS						
Aluminum	300 - 400	91 - 122	7	8	30	20
Brass (leaded)	300 - 600	91 - 182	6	7	0	0
Bronze (free-cutting)	300 - 600	91 - 182	6	5	0	2
Bronze (tough)	80 - 150	24 - 46	12	15	15	25
Copper	80 - 150	24 - 46	7	7	10	25
Magnesium alloy	300 - 400	91 - 122	8	8	6	4

NOTE: SPEEDS IN m/min ARE THE APPROXIMATE EQUIVALENT OF ft/min. FC = FRONT CLEARANCE; SC = SIDE CLEARANCE; BR = BACK RAKE; SR = SIDE RAKE ANGLE. (SEE FIGURE 3-54 FOR CLEARANCE AND RAKE.)

Figure 3-52. Recommended cutting speed and cutting tool angles for turning.

PROCESS	FEED		DEPTH-OF-CUT	
	ipr	ipr	in	mm
Rough cutting	.010 - .030	0.25 - 0.75	.040 - .150	.1 - 3.8
Finishing	.003 - .005	0.08 - 0.13	.005 - .030	0.13 - 0.76

Figure 3-53. Recommended cutting feed and depth of cut for turning and boring.

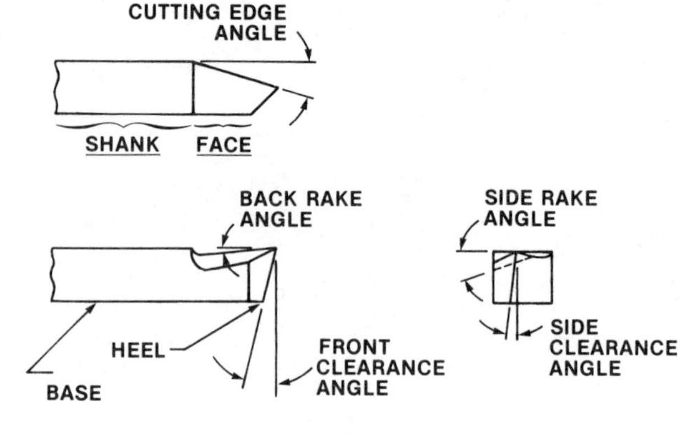

Figure 3-54. Relative position of tool bit angles.

Taper Turning. A taper is a solid or hollow cylinder whose diameter increases or decreases uniformly from one end to the other. The difference between the large diameter and the small diameter is called the *taper*, and is expressed as taper-per-foot (Tpf), or taper-per-inch (Tpi). See Figure 3-57 for the components of a taper. The components of any taper are related to one another and can be computed. The following taper formulas can be used to calculate any unknown component:

1. $Tpi = \frac{D - d}{l}$ or $\frac{Tpf}{12}$ — Tpi = taper-per-inch
2. $Tpf = 12\ Tpi$ — Tpf = taper-per-foot
3. $\tan \frac{a}{2} = \frac{D - d}{2 \times l}$ or $\frac{Tpi}{2}$ — $\frac{a}{2}$ = angle with the center line of taper
4. $D = (Tpi \times l) + d$ — D = large diameter of taper
5. $d = D - (l \times Tpi)$ — d = small diameter of taper
6. $l = \frac{D - d}{Tpi}$ — l = length of taper

See Example 3-5 for an application of a typical problem.

Example 3-5: The large diameter of a 4″ long taper is 1.750″ and the small diameter is 1.250″. Determine the taper-per-inch (Tpi), the taper-per-foot (Tpf), and the included angle (a) of this taper.

Given: l = 4″, D = 1.750″, d = 1.250″

Solution: $Tpi = \frac{D - d}{l} = \frac{1.750 - 1.250}{4} = \frac{0.500}{4} = 0.125''$

$$Tpf = 12 \times Tpi = 12 \times 0.125 = 1.5''$$

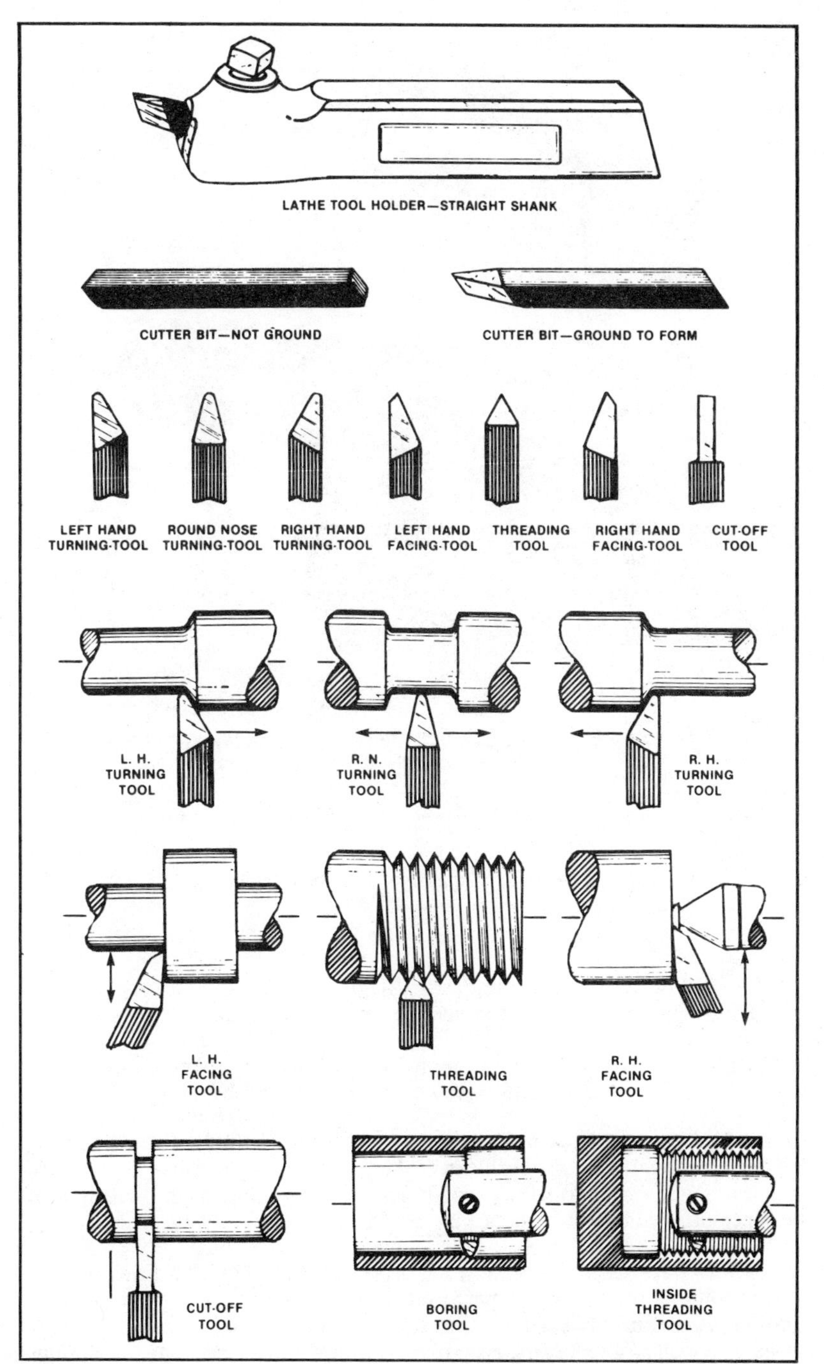

Figure 3-55. Common lathe tool bit shapes and applications. (South Bend Lathe, Inc.)

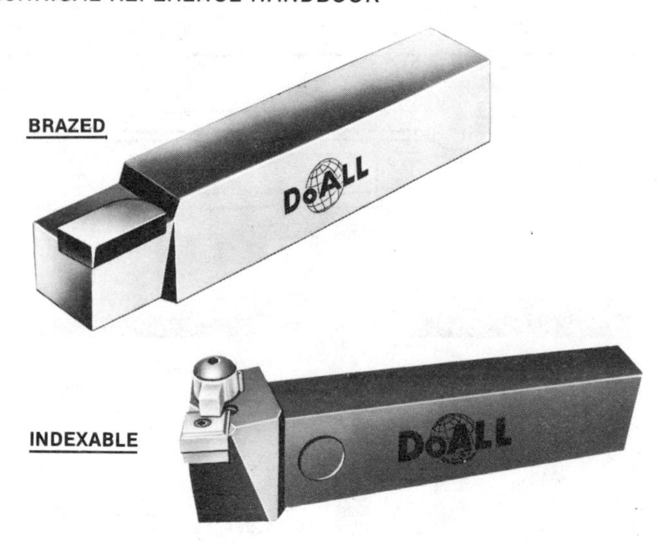

Figure 3-56. Carbide tipped tools and holders. (DoAll Company)

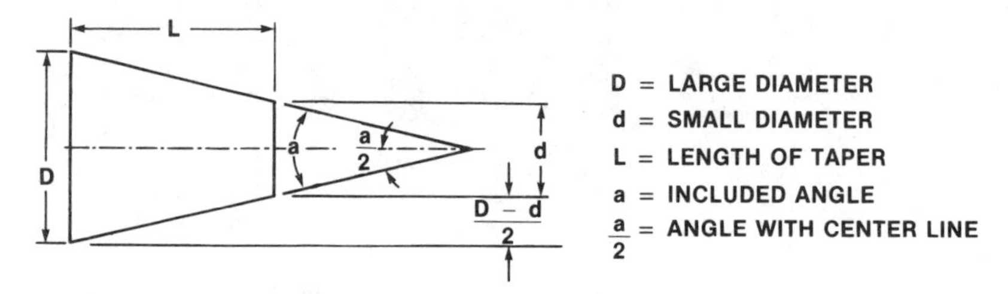

Figure 3-57. Taper and its components.

Three methods of taper turning with a single-point cutting tool are the *compound rest method, taper attachment method,* and *tailstock offset method.*

Compound Rest Method. The compound rest method of taper turning (Figure 3-58) is suitable for short and steep, external and internal tapers.

Taper Attachment Method. The taper attachment method of taper turning (Figure 3-59) is suitable for both external and internal tapers with an included angle of 20° or less, and with a length equal to or smaller than the length of the taper attachment. It also can be used effectively to cut tapered screws.

Tailstock Offset Method. The tailstock offset method of taper turning (Figure 3-60) is suitable for external tapers only. It is a less precise method than the compound rest and taper attachment methods, but it can be used effectively for short as well as long tapers. Example 3-6 shows how to calculate the amount of tailstock offset (x in Figure 3-60).

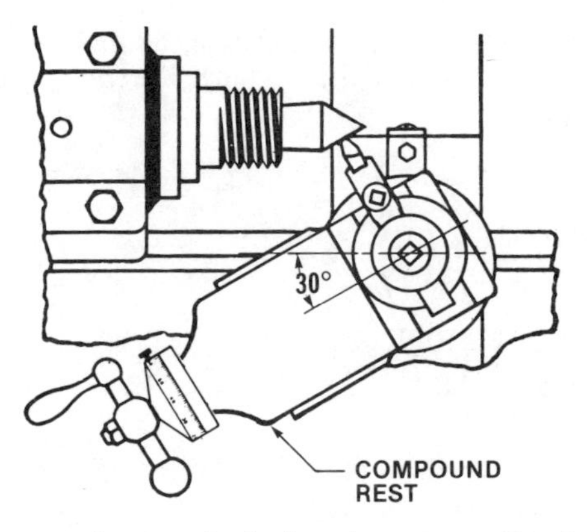

Figure 3-58. Compound rest method of turning a taper. (South Bend Lathe, Inc.)

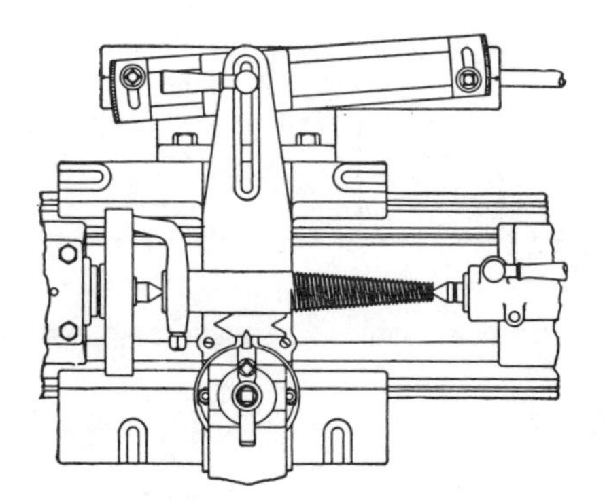

Figure 3-59. Taper attachment method of turning a taper. (South Bend Lathe, Inc.)

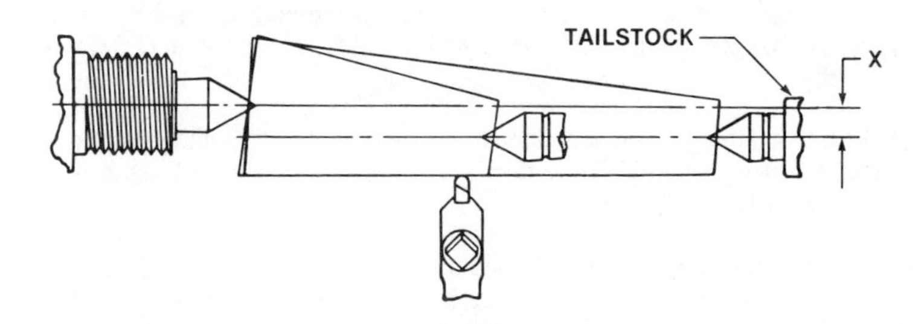

Figure 3-60. Tailstock offset method of turning a taper. (South Bend Lathe, Inc.)

Example 3-6: A taper 8″ long with a large diameter of 2″ and a small diameter of 1.75″, is to be made on a shaft 10″ long. Determine the amount of the tailstock offset (x).

Given: l = 8″, D = 2″, d = 1.75″, L = 10″

Solution: $x = \left(\frac{D - d}{2}\right)\left(\frac{L}{1}\right) = \left(\frac{2.00 - 1.75}{2}\right)\left(\frac{10}{8}\right)$

$= \left(\frac{0.25}{2}\right)\left(\frac{10}{8}\right) = \frac{2.5}{16} = \frac{5''}{32}$

Standard Tapers. The two kinds of standard tapers (Figure 3-61) are the *self-holding tapers* and *self-releasing tapers.*

TYPE OF TAPER	Tpf	Tpi	ANGLE a	ANGLE a/2	USES
*Morse	5/8″	.0521″	3°	1° 30′	Lathe centers, drill shanks, and collets
*Brown & Sharpe	1/2″	.04166″	2° 23′	1° 11′	Milling machine spindles and milling cutter shanks
*Jarno or Metric	.600″	.050″	2° 52′	1° 26′	Various types of lathes and drill presses
*American National Standard	3/4″	.0625″	3° 35′	1° 48′	Special machine tools
†Milling Machine Arbor	3 1/2″	.2916″	16° 36′	8° 18′	Milling Machine Arbors and Spindles

* Self-holding taper
† Self-releasing taper

NOTE: THE METRIC TAPER HAS AN INCLUDED ANGLE SIMILAR TO THE JARNO; HOWEVER, ITS DIMENSIONS ARE IN METRIC UNITS.

Figure 3-61. Characteristics and uses of standard tapers.

Self-holding Tapers. The self-holding tapers have an included angle ranging from 2° 23′ to 3° 35′, and taper-per-foot of .500″ to .750″. Each shank is firmly seated in its socket, thus providing considerable resistance to any force turning it in the socket. These tapers are used in taper-shank tools and machine parts such as twist drills, drill chucks, reamers, lathe centers, and end mills.

The four types of self-holding tapers, each of which is available in a variety of sizes, are *Morse, Brown & Sharpe, Jarno* or *Metric,* and *American National Standard.*

Self-releasing Tapers. The self-releasing tapers have an included angle of 16° 36′ and a Tpf of 3.500″. These tapers are used mainly in milling and boring machine spindles.

NOTE: The taper-per-foot of Morse, and Brown & Sharpe is approximately 5/8″ and 1/2″, respectively. Figure 3-62 shows the exact dimensions of the standard Morse tapers. See Figure 3-63 for the milling machine arbor tapers dimensions.

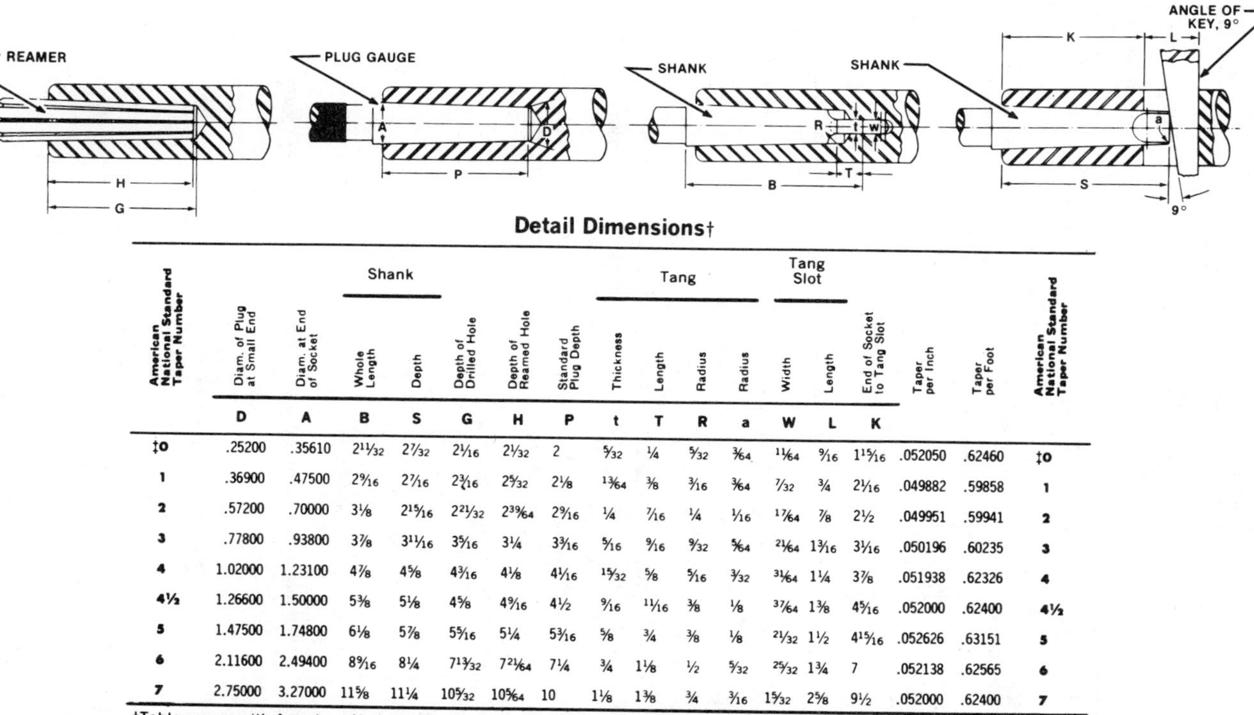

Detail Dimensions†

American National Standard Taper Number	Diam. of Plug at Small End	Diam. at End of Socket	Shank: Whole Length	Shank: Depth	Depth of Drilled Hole	Depth of Reamed Hole	Standard Plug Depth	Tang: Thickness	Tang: Length	Tang: Radius	Tang: Radius	Tang Slot: Width	Tang Slot: Length	End of Socket to Tang Slot	Taper per Inch	Taper per Foot	American National Standard Taper Number
	D	A	B	S	G	H	P	t	T	R	a	W	L	K			
‡0	.25200	.35610	2 11/32	2 7/32	2 1/16	2 1/32	2	5/32	1/4	5/32	3/64	11/64	9/16	1 15/16	.052050	.62460	‡0
1	.36900	.47500	2 9/16	2 7/16	2 3/16	2 5/32	2 1/8	13/64	3/8	3/16	3/64	7/32	3/4	2 1/16	.049882	.59858	1
2	.57200	.70000	3 1/8	2 15/16	2 21/32	2 39/64	2 9/16	1/4	7/16	1/4	1/16	17/64	7/8	2 1/2	.049951	.59941	2
3	.77800	.93800	3 7/8	3 11/16	3 5/16	3 1/4	3 3/16	5/16	9/16	9/32	5/64	21/64	1 3/16	3 1/16	.050196	.60235	3
4	1.02000	1.23100	4 7/8	4 5/8	4 3/16	4 1/8	4 1/16	15/32	5/8	5/16	3/32	31/64	1 1/4	3 7/8	.051938	.62326	4
4½	1.26600	1.50000	5 3/8	5 1/8	4 5/8	4 9/16	4 1/2	9/16	11/16	3/8	1/8	37/64	1 3/8	4 5/16	.052000	.62400	4½
5	1.47500	1.74800	6 1/8	5 7/8	5 5/16	5 1/4	5 3/16	5/8	3/4	3/8	1/8	21/32	1 1/2	4 15/16	.052626	.63151	5
6	2.11600	2.49400	8 9/16	8 1/4	7 13/32	7 21/64	7 1/4	3/4	1 1/8	1/2	5/32	25/32	1 3/4	7	.052138	.62565	6
7	2.75000	3.27000	11 5/8	11 1/4	10 5/32	10 5/64	10	1 1/8	1 3/8	3/4	3/16	1 5/32	2 5/8	9 1/2	.052000	.62400	7

†Table agrees with American National Standards for Taper Shanks except for angle and undercut of tang.
‡Size 0 taper shank not listed in American National Standards.

Figure 3-62. Essential dimensions of Morse tapers. (Cleveland Twist Drill Company)

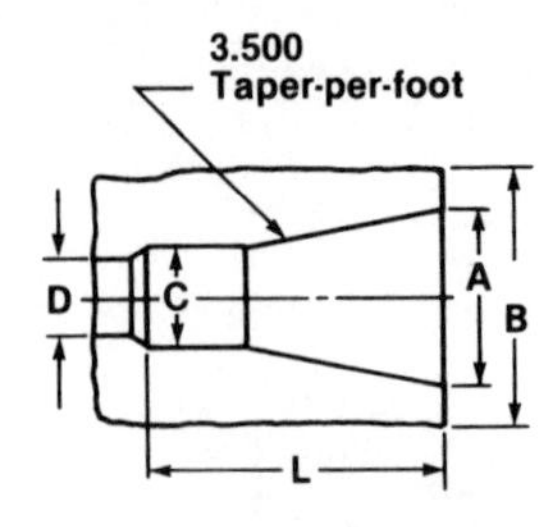

INSIDE TAPER

TAPER NO.	A	B	C	D	L	N
30	1.250	2.7493	.692 .685	21/32	2 7/8	1.250
40	1.750	3.4993	1.005 .997	21/32	3 7/8	1.750
50	2.750	5.0618	1.568 1.559	1 1/16	5 1/2	2.750
60	4.250	8.718	2.381 2.371	1 3/8	8 5/8	4.250

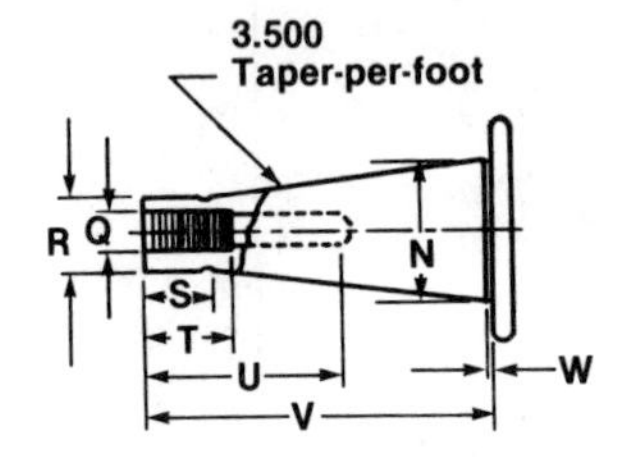

OUTSIDE TAPER

TAPER NO.	Q	R	S	T	U	V	W
30	1/2 – 13	.675 .673	13/16	1	2	2 3/4	1/16
40	5/8 – 11	.987 .985	1	1 1/8	2 5/16	3 3/4	1/16
50	1 – 8	1.549 1.547	1	1 3/4	3 1/2	5 1/8	1/8
60	1 1/4 – 7	2.361 2.359	1 3/4	2 1/4	4 1/4	8 5/16	1/8

Figure 3-63. Essential dimensions of standard milling machine arbor tapers.

Thread Cutting. Thread cutting is the process of producing a helical, or spiral groove on the surface of solid or hollow cylindrical stock. A single-edge cutting tool with exactly the same shape as the desired thread is used. In order to cut an American National or Unified National screw thread, the cutting tool must be ground properly to form a 60° angle. The accuracy of this angle is checked with the center gauge (Figure 3-64). The center gauge is also used for setting the cutting tool for external threading (Figure 3-65) and internal threading (Figure 3-66).

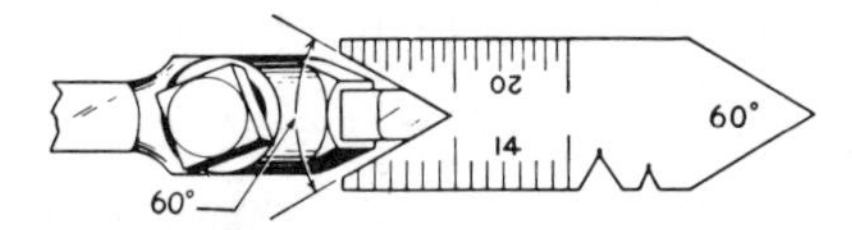

Figure 3-64. Center gauge used for checking proper cutting angle. (South Bend Lathe, Inc.)

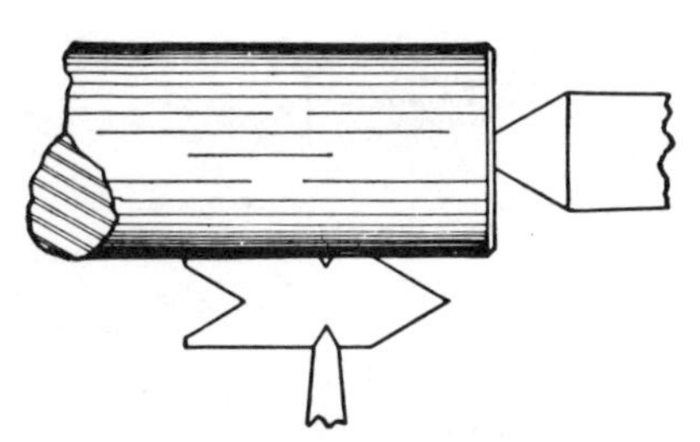

Figure 3-65. Using the center gauge to set the cutting tool for external thread cutting. (South Bend Lathe, Inc.)

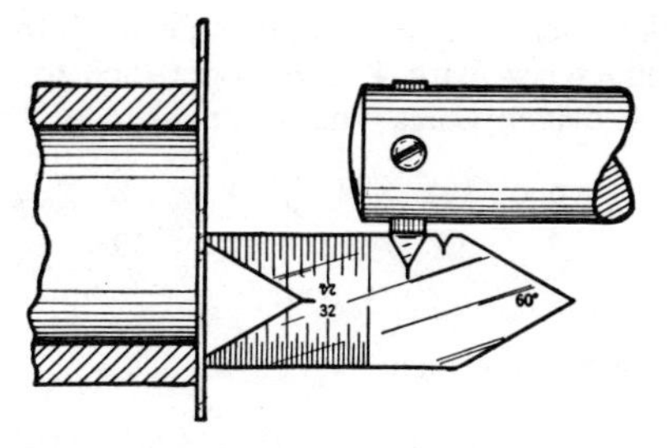

Figure 3-66. Using the center gauge to set the cutting tool for internal thread cutting. (South Bend Lathe, Inc.)

Spindle and Lead Screw Gear Ratio for Thread Cutting. In order to cut a thread, the tool must move at every turn of the spindle a distance equal to the pitch of the thread. This movement is accomplished through a train of gears from the spindle (drive gear) to the lead screw (driven gear).

The ratio between the number of teeth of the drive gear (G_1) and the driven gear (G_2) is equal to the ratio of the pitch of the thread to be cut (p) and the pitch of the lead screw (P). That is,

$$G_1/G_2 = p/P.$$

Whenever the ratio p/P is greater than $1/6$, a *compound gearing* (Figure 3-67) must be used; that is, instead of one pair of gears, two pairs of gears are used.

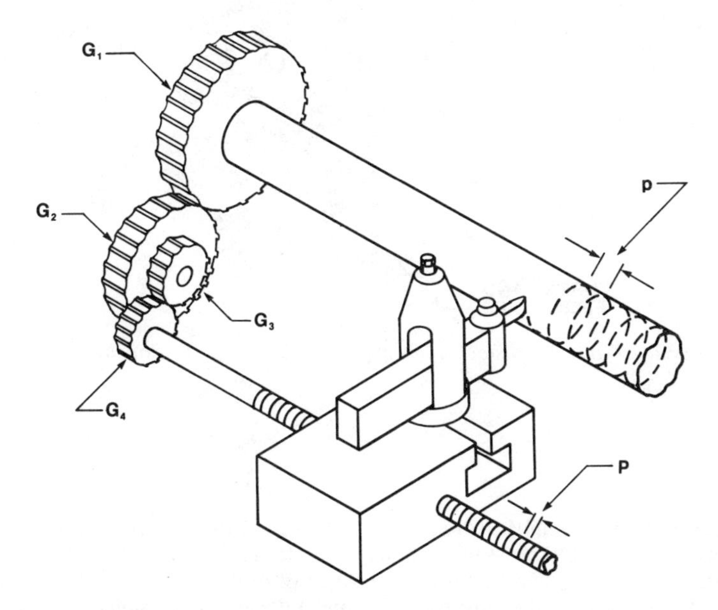

Figure 3-67. Relative position of gears in compound gearing.

For example, to make a screw with 32 threads-per-inch [pitch (p) = 1/32″] on a lathe that has a lead screw with 4 threads-per-inch [pitch (P) = 1/4″], the ratio of the drive and driven gears would be:

$$\frac{G_1}{G_2} = \frac{p}{P} = \frac{1/32}{1/4} = \frac{1}{32} \times \frac{4}{1} = \frac{4}{32} = \frac{1}{8}$$

The ratio 1/8 can be analyzed as:

$$\frac{1}{8} = \frac{1}{4} \times \frac{1}{2} \quad \text{or} \quad \frac{p}{P} = \frac{G_1}{G_2} \times \frac{G_3}{G_4}$$

Modern lathes are equipped with a gear box containing a number of gears to provide a larger number of ratios between the spindle and the lead screw. A thread and feed chart attached to the gear box of the headstock (Figure 3-68) indicates the positions of the sliding gear handle, the thread and feed selector handle, and the selector knob for cutting threads with the desired number of threads-per-inch.

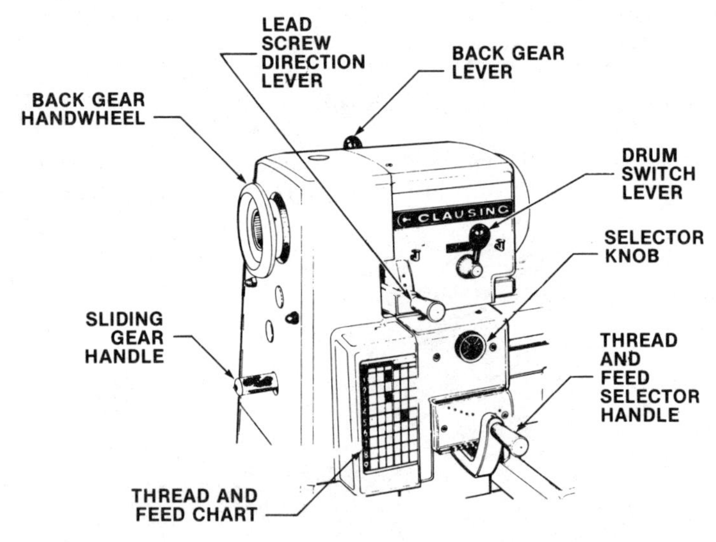

Figure 3-68. Typical headstock nomenclature. (Clausing Division of Atlas Press Company)

Suggestions for Thread Cutting. In thread cutting, the tool post slide on the carriage (Figure 3-69) must be set at a 29° angle. The full depth of any thread is obtained in successive cuts (Figure 3-70). For each successive cut, the half-nut lever is engaged at the same time that one of the graduations on the threading dial indicator coincides with the fixed point (Figure 3-71).

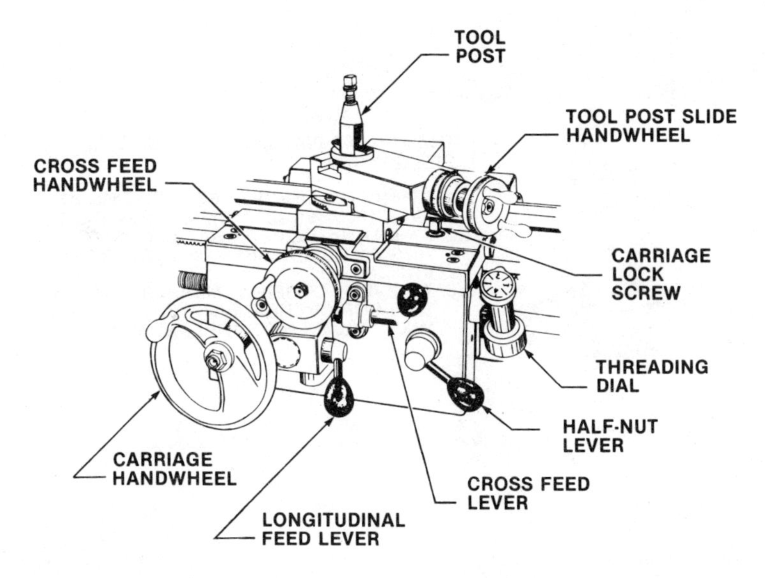

Figure 3-69. Typical carriage nomenclature. (Clausing Division of Atlas Press Company)

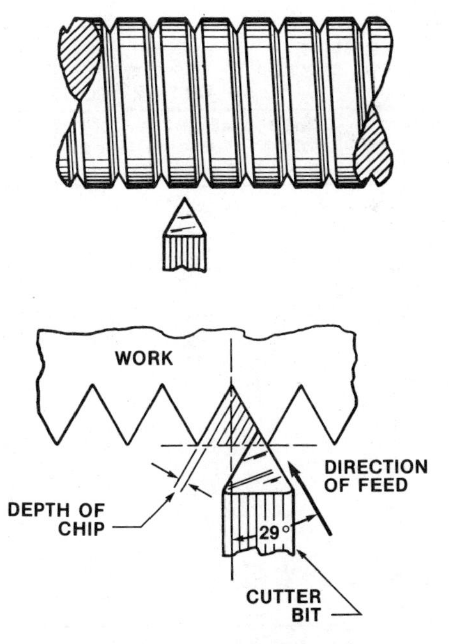

Figure 3-70. Successive cuts in thread cutting. (South Bend Lathe, Inc.)

THE THREAD DIAL INDICATOR SHOULD BE ENGAGED:

1. AT ANY GRADUATION FOR CUTTING AN EVEN NUMBER OF THREADS PER INCH.
2. AT ANY NUMBERED GRADUATION FOR CUTTING AN ODD NUMBER OF THREADS PER INCH.
3. AT EVERY OTHER NUMBERED GRADUATION FOR CUTTING FRACTIONAL OR A MIXED NUMBER OF THREADS PER INCH.

Figure 3-71. Thread dial indicator. (South Bend Lathe, Inc.)

Depth-of-Thread. The depth of a thread is equal to one-half the difference between the thread's major and minor diameter; that is,

$$\text{Depth-of-thread} = \frac{\text{Major diameter} - \text{Minor diameter}}{2}$$

Successful thread cutting depends, to a great extent, on the correct depth-of-thread (Figure 3-72).

		WHEN CUT WITH NATIONAL FORM TOOL		WHEN CUT WITH 60° V-TYPE VEE FORM TOOL		DEPTH OF COMPOUND FEED REST SET AT 29° SINGLE DEPTH	
Threads-per-inch	Pitch Inches	Single depth-of-thread	Double depth-of-thread	Single depth-of-thread	Double depth-of-thread	N. F. Tool	Vee Form Tool
4	.2500	.1624	.3248	.1894	.3789	.186	.216
4 1/2	.2222	.1443	.2887	.1684	.3368	.165	.193
5	.2000	.1299	.2598	.1516	.3031	.148	.173
5 1/2	.1818	.1181	.2362	.1378	.2755	.135	.157
6	.1667	.1083	.2165	.1263	.2525	.124	.144
7	.1429	.0928	.1856	.1082	.2165	.106	.123
8	.1250	.0812	.1624	.0947	.1894	.093	.108
9	.1111	.0722	.1443	.0842	.1684	.083	.095
10	.1000	.0650	.1299	.0758	.1515	.074	.087
11	.0909	.0590	.1181	.0689	.1377	.067	.078
12	.0833	.0541	.1083	.0631	.1263	.062	.072
13	.0769	.0500	.0999	.0583	.1166	.057	.067
14	.0714	.0464	.0928	.0541	.1082	.053	.062
16	.0625	.0406	.0812	.0473	.0947	.046	.054
18	.0556	.0361	.0722	.0421	.0842	.041	.047
20	.0500	.0325	.0650	.0379	.0758	.037	.043

Figure 3-72. Partial listing of proper depth-of-thread for various standard threads. (Clausing Division of Atlas Press Company)

Pitch Diameter of a Thread. The *pitch diameter* of a thread is the diameter of an imaginary circle passing through the threads at points that make the width of the threads and the width of the spaces (grooves) equal. The pitch diameter of a thread can be determined with a screw thread micrometer (Figure 3-73), or by using the *three-wire method* (Figure 3-74). Also available are manufacturer's tables that furnish thread pitch diameters (Figures 3-116 and 3-117).

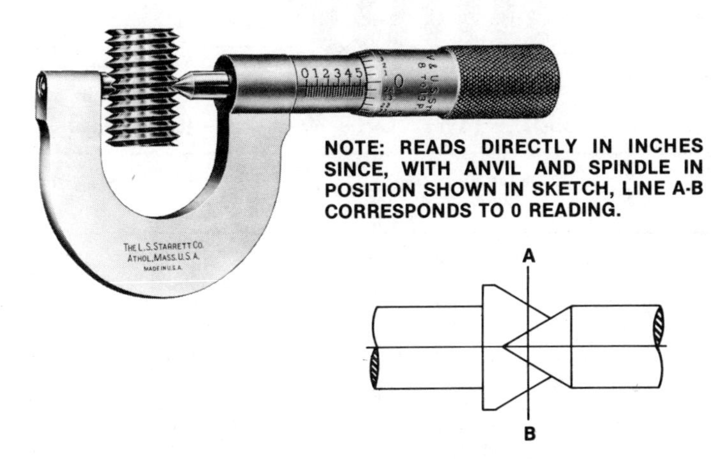

Figure 3-73. Measuring pitch diameter with a screw thread micrometer. (The L. S. Starrett Company)

Three-Wire Method. The three-wire method of determining the pitch diameter of a thread uses three precision wires, each having the same diameter. This method is used when measuring the pitch diameter of precision threads. By measuring over the wires with a micrometer and taking the measurement M, the pitch diameter E can be found by applying the formulas in Figure 3-74. See Example 3-6A.

Example 3-6A: Find the measurement (M) over the wires and the pitch diameter (E) of a standard UN 3/4 - 10 thread.

Given: Major diameter D = 0.750″, Pitch p = 0.100″, Best wire size W = 0.05774″

Solution:
$$\begin{aligned} M &= D - (1.5155p) + 3W \\ &= 0.750 - (1.5155 \times 0.100) + (3 \times 0.05774) \\ &= 0.750 - 0.15155 + 0.17322 = 0.59845 + 0.17322 \\ &= 0.77167'' \\ M &= E - (0.88603p) + 3W \quad \text{or} \\ &= M + (0.88603p) - 3W \\ &= 0.77167 + (0.86603 \times 0.100) - (3 \times 0.05774) \\ &= 0.77167 + 0.086603 - 0.17322 = 0.85827 - 0.17322 \\ &= 0.685'' \end{aligned}$$

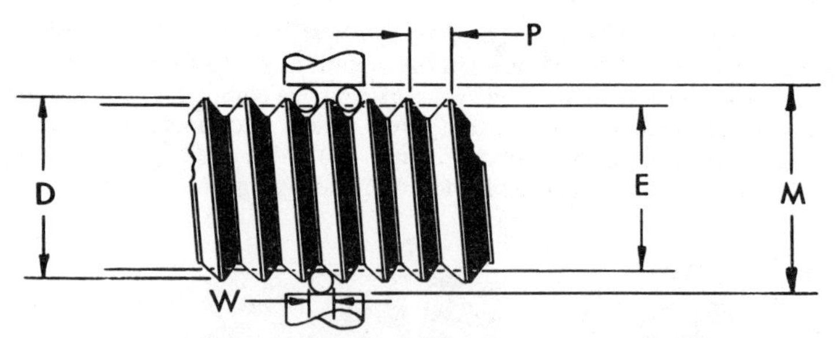

$$M = D - (1.5155p) + (3W) \quad \text{OR} \quad M = E - (0.86603p) + 3W$$

Threads per Inch n	Pitch $p = \frac{1}{n}$	Best Wire Size 0.577350 p	Threads per Inch n	Pitch $p = \frac{1}{n}$	Best Wire Size 0.577350 p
80	.012500	.00722	18	.055556	.03208
72	.013889	.00802	16	.062500	.03608
64	.015625	.00902	14	.071429	.04124
56	.017857	.01031	13	.076923	.04441
50	.020000	.01155	12	.083333	.04811
48	.020833	.01203	11½	.086957	.05020
44	.022727	.01312	11	.090909	.05249
40	.025000	.01443	10	.100000	.05774
36	.027778	.01604	9	.111111	.06415
32	.031250	.01804	8	.125000	.07217
30	.033333	.01924	7½	.133333	.07698
28	.035714	.02062	7	.142857	.08248
27	.037037	.02138	6	.166667	.09623
26	.038462	.02221	5½	.181818	.10497
24	.041667	.02406	5	.200000	.11547
22	.045445	.02624	4½	.222222	.12830
20	.050000	.02887	4	.250000	.14434

Figure 3-74. Formulas used to determine pitch diameter with the three-wire method. (Clausing Division of Atlas Press Company)

Boring Machine

The basic machine tool used for boring is the boring machine, or boring mill. The two types of boring machines are the *horizontal boring machine* (Figure 3-75) and the *vertical boring machine* (Figure 3-76).

In the horizontal boring machine (Figure 3-77), the workpiece is stationary and the cutting tool rotates. In the vertical boring machine (Figure 3-78), the cutting tool is stationary and the workpiece rotates.

NOTE: The cutting tool angles, as well as the recommended cutting speed and feed used for boring, are approximately the same as those used for lathe turning.

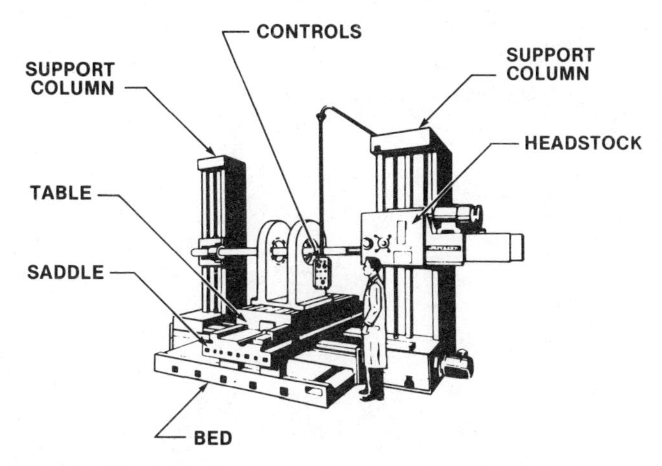

Figure 3-75. Basic horizontal boring machine, or boring mill, nomenclature. (Giddings and Lewis Machine Tool Co.)

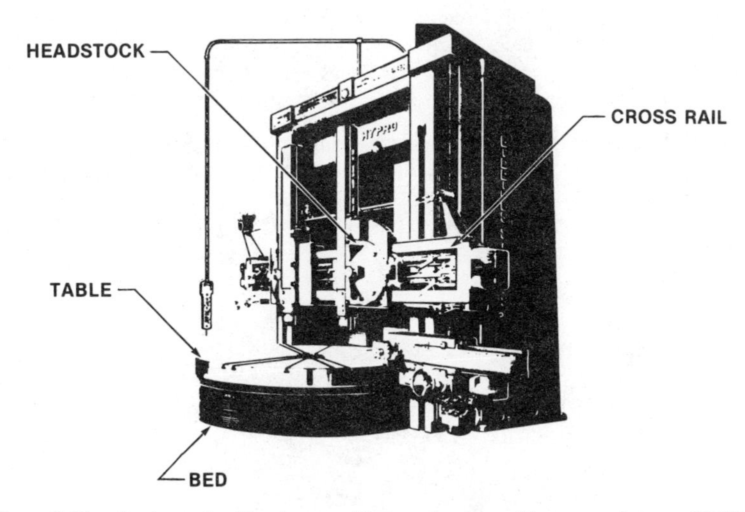

Figure 3-76. Basic vertical boring machine, or boring mill, nomenclature. (Giddings and Lewis Machine Tool Co.)

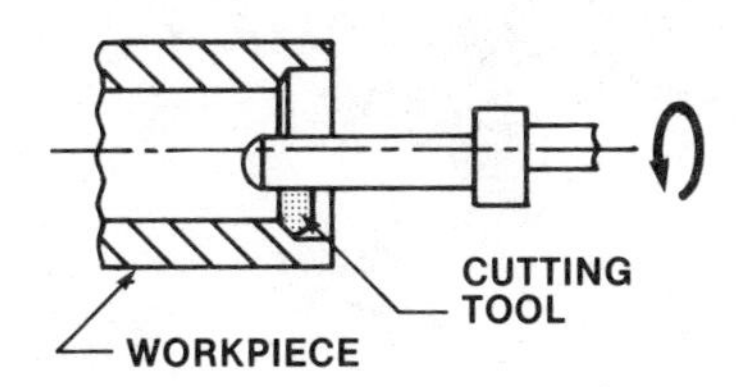

Figure 3-77. Horizontal boring tool action (tool rotates and workpiece is stationary).

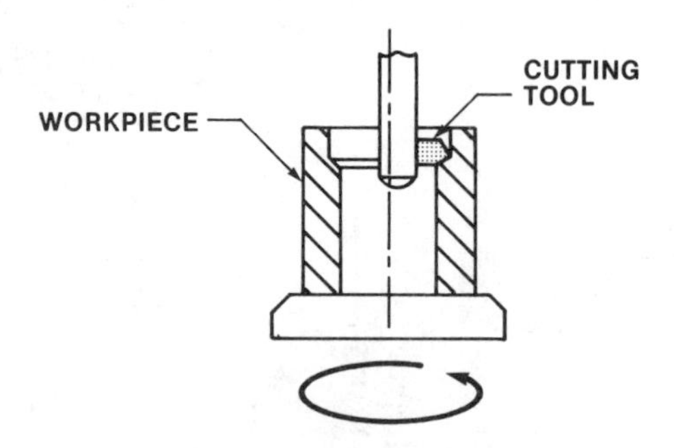

Figure 3-78. Vertical boring tool action (table with workpiece rotates and tool is stationary).

Shaper and Planer

In the shaper (Figure 3-79), a reciprocating ram provides the cutting speed and the table provides the feed. In the planer (Figure 3-80), a reciprocating table provides the cutting speed and the head provides the feed. See Figure 3-81 for recommended cutting speeds and feeds for shaping and planing.

Number of Strokes and Machining Time for Shaping and Planing. The number of strokes-per-minute of any shaping or planing operation depends upon the recommended cutting speed, the length of the stroke, and a constant. The value of this constant is approximately 0.70 for shapers with mechanical drive and

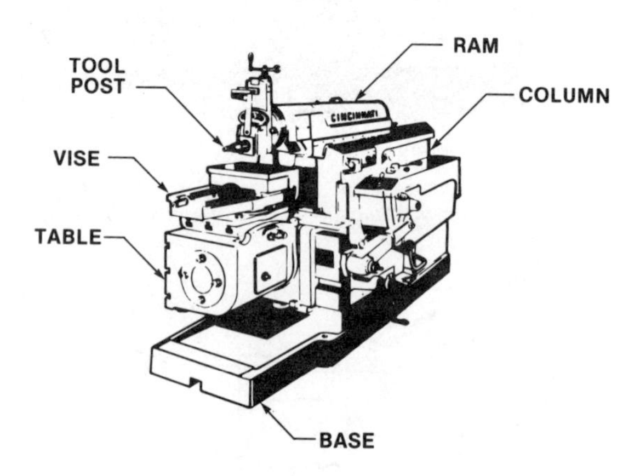

Figure 3-79. Basic shaper nomenclature. (Cincinnati Incorporated)

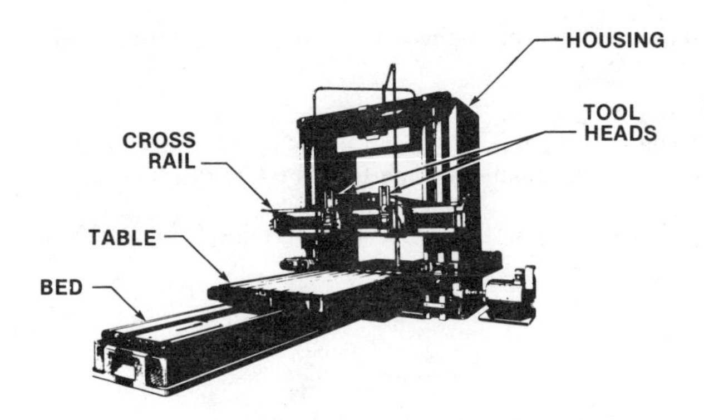

Figure 3-80. Basic planer nomenclature. (Giddings and Lewis Machine Tool Co.)

MATERIAL	HIGH-SPEED STEEL TOOL				CARBIDE TOOL			
	CUTTING SPEED		FEED		CUTTING SPEED		FEED	
	ft/min	m/min	in/st	mm/st	ft/min	m/min	in/st	mm/st
Carbon steel	60 - 80	18 - 24	.010	0.25	140 - 160	42 - 43	.010	0.25
Alloy steel	30 - 50	9 - 15	.015	0.38	140 - 150	42 - 45	.012	0.30
Cast iron	40 - 60	12 - 13	.020	0.50	80 - 100	24 - 30	.012	0.30
Aluminum alloy	150 - 200	45 - 60	.010	0.25	200 - 300	60 - 90	.015	0.38
Copper alloy	150 - 200	45 - 60	.010	0.25	200 - 300	60 - 90	.015	0.38

Figure 3-81. Recommended cutting speed and feed for shaping and planing.

up to 0.90 for shapers with hydraulic drive. The following formulas can be used to calculate the number of strokes-per-minute:

$$N = \frac{12 \times CS}{L + 1} (c)$$

N = number of strokes-per-minute (strk/min)
CS = cutting speed in feet-per-minute (ft/min)
L = length of workpiece in inches
c = 0.70 − 0.90

$$N = \frac{1{,}000 \times V}{L + 25} (c)$$

N = number of strokes-per-minute (strk/min)
V = cutting speed in meters-per-minute (m/min)
L = length of workpiece in millimeters (mm)
c = 0.70 − 0.90

See Example 3-7 for an application of computing strokes-per-minute and machining time.

Example 3-7: A cast-iron block 20″ long by 12″ wide is to be machined in a shaper. Determine the time required for one pass if the recommended cutting speed is 60 feet-per-minute and the feed is 0.020 inch-per-stroke.

Given: $L = 20''$, $W = 12''$, $CS = 60$ ft/min, $f = 0.020$ in/st

$$\text{Solution: } N = \left(\frac{12 \times CS}{L + 1}\right)(c) = \left(\frac{12 \times 60}{20 + 1}\right) 0.7 = \frac{12 \times 60 \times 0.7}{21}$$

$$= \frac{504}{21} = 24 \text{ strk/min}$$

$$t = \frac{W}{N \times f} = \frac{12.0}{24 \times 0.020} = \frac{12}{0.48} = 25 \text{ min}$$

Cutting Tools for Shaper and Planer. Shapers and planers use single-edge cutting tools (Figure 3-82).

Cutting Tools Recommended for MILD STEEL

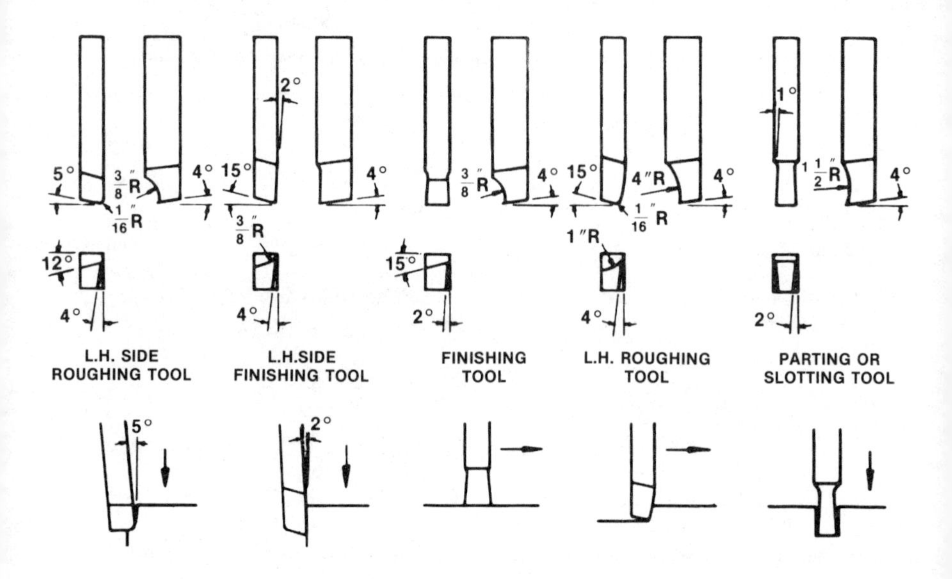

continued

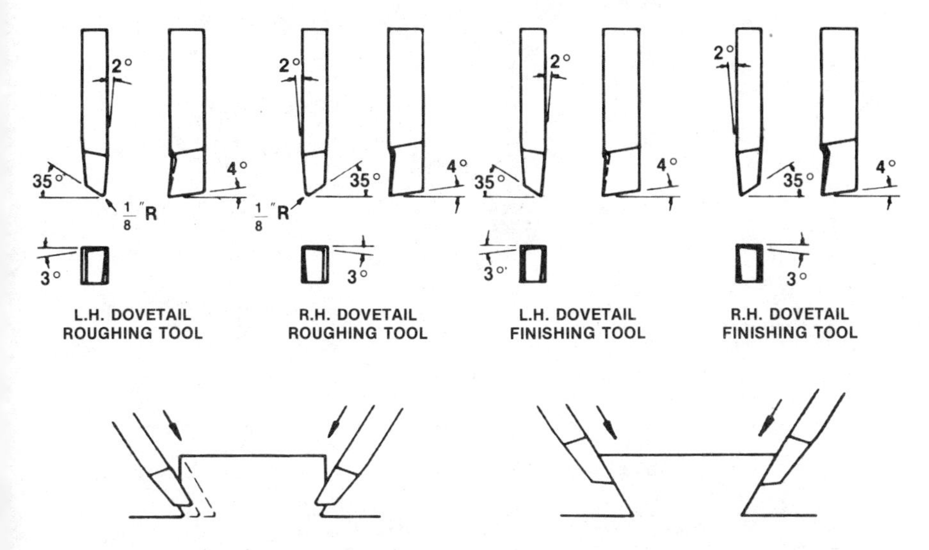

Figure 3-82. Typical shaper-planer cutting tools. (Cincinnati Incorporated)

Milling Machine

The two types of milling machines are the *horizontal milling machine* (Figure 3-83) and the *vertical milling machine* (Figure 3-84).

In both the horizontal and vertical milling machines, cutting takes place when the cutting tool is rotating and the table is providing the feed. Figure 3-85 shows recommended cutting speeds and feeds for milling cutters. Milling cutters are multi-edge cutting tools. Each type of milling cutter is characterized by its style

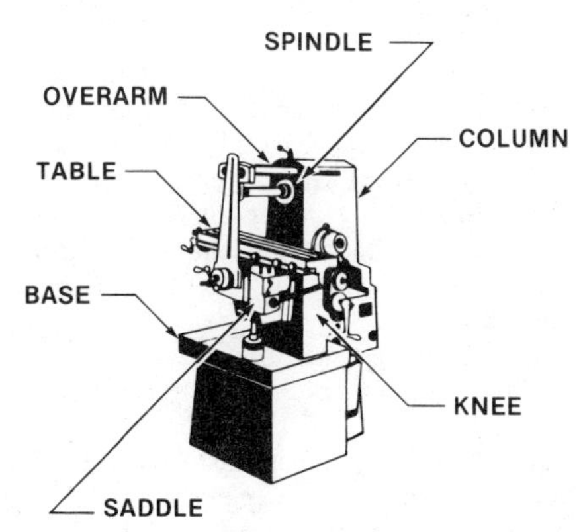

Figure 3-83. Basic horizontal milling machine nomenclature. (Clausing Division of Atlas Press Company)

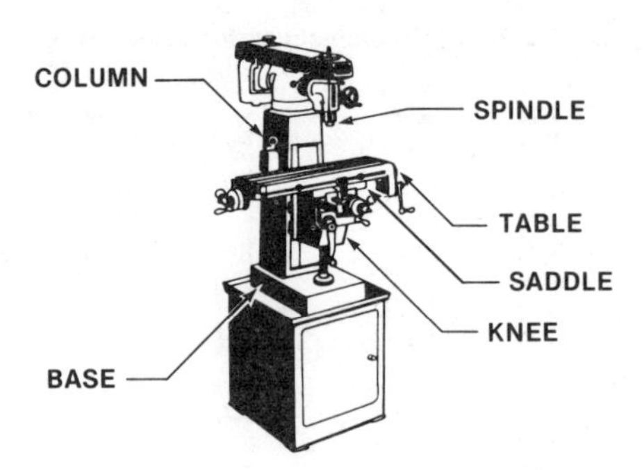

Figure 3-84. Basic vertical milling machine nomenclature. (Clausing Division of Atlas Press Company)

MATERIAL	CUTTING SPEED		FACE MILL CUTTER		SPIRAL MILL CUTTER		SIDE MILL CUTTER		END MILL CUTTER	
			FEED-PER-TOOTH							
	ft/min	m/min	ipt	mmpt	ipt	mmpt	ipt	mmpt	ipt	mmpt
Low-carbon steel	100 - 120	30 - 36	.011	0.28	.009	0.23	.007	0.18	.005	0.13
Medium-carbon steel	80 - 100	24 - 30	.009	0.23	.007	0.18	.006	0.15	.005	0.13
High-carbon steel	60 - 80	18 - 24	.009	0.23	.007	0.18	.006	0.15	.005	0.13
Alloy steel (soft)	50 - 60	15 - 18	.008	0.20	.006	0.15	.005	0.13	.004	0.10
Alloy steel (hard)	40 - 50	12 - 15	.005	0.13	.004	0.10	.003	0.08	.003	0.08
Cast iron (soft)	50 - 60	15 - 18	.018	0.46	.012	0.30	.011	0.28	.009	0.23
Cast iron (hard)	80 - 100	18 - 30	.015	0.38	.010	0.25	.009	0.23	.008	0.20
Aluminum alloys	200 - 400	60 - 120	.022	0.56	.018	0.46	.014	0.35	.011	0.28
Copper alloys	150 - 250	50 - 75	.020	0.51	.017	0.43	0.13	0.33	.010	0.25

Figure 3-85. Recommended cutting speed and feed for milling.

and shape, cutter diameter, hole diameter or taper shank size, and width. Milling cutters are made of high-speed steel, cast alloy inserts, and cemented-carbide tips. Figure 3-86 shows typical milling cutters.

For effective milling, the milling cutter must be properly ground to form sharp teeth. Each tooth of a milling cutter is characterized by tooth depth, land, flute, and various angles (Figure 3-87). The function of these angles is to decrease friction and facilitate cutting.

Cemented-carbide milling cutters use interchangeable carbide inserts. Using cemented-carbide milling cutters results in higher productivity and lower operating costs.

In milling machines, the feeding mechanism provides for a variety of settings and is independent of the spindle speed. Each of these settings moves the table at a specific rate of feed-per-minute. This arrangement enables the operator to use faster rates of feed for larger, slowly rotating milling cutters that have many cutting edges.

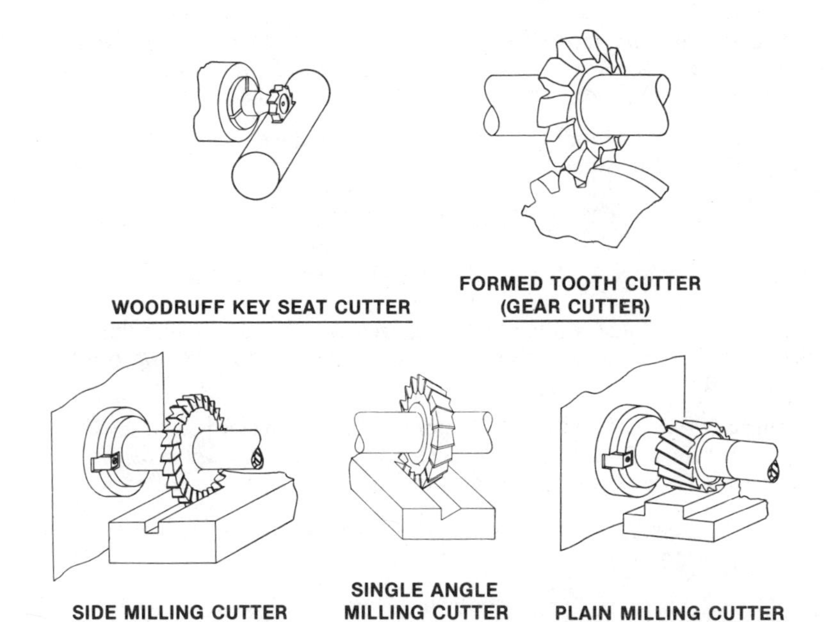

Figure 3-86. Common milling cutters. (National Twist Drill-Division of Lear Siegler, Inc.)

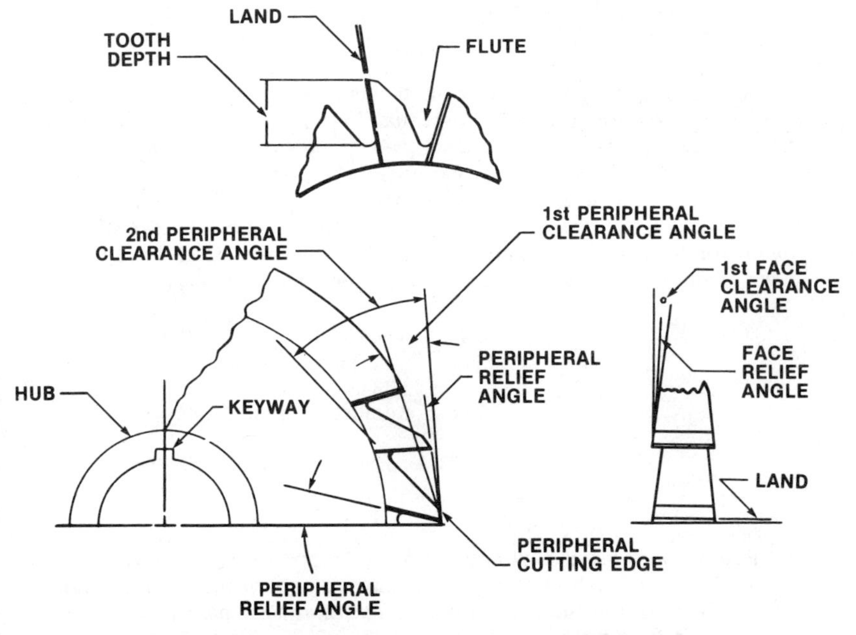

Figure 3-87. Milling cutter nomenclature. (DoAll Company)

In cases where the feed is given in inches-per-tooth or millimeters-per-tooth, the operator must convert the feed-per-tooth into feed-per-minute to determine the proper setting. See Example 3-8.

Example 3-8: A 12-tooth milling cutter is rotating with 48 RPM. Determine the rate of feed-per-minute (F) of this cutter if the feed-per-tooth is 0.004″.

Given: $n = 12$, $N = 48$ RPM, $f = 0.004$ ipt

Solution: $F = n \times f \times N = 12 \times 0.004 \times 48 = 2.3$ in/min

Machining Time for Milling. The time required for any milling operation depends upon the rate of feed-per-minute and the length of the workpiece. See Example 3-9.

Example 3-9: A 20-tooth milling cutter with a 4″ diameter is used to machine a piece 15″ long. Determine the time required for one pass if the recommended speed is 80 feet-per-minute and the feed is 0.0025 inch-per-tooth.

Given: $D = 4''$, $CS = 80$ ft/min, $L = 15''$, $f = 0.0025$ ipt,
$n = 20$ teeth

Solution: $$N = \frac{12 \times CS}{\pi \times D} = \frac{12 \times 80}{3.14 \times 4} = \frac{960}{12.56} = 76 \text{ RPM}$$

$$F = n \times f \times N = 20 \times 0.0025 \times 76 = 3.8 \text{ in/min}$$

$$t = \frac{L}{F} = \frac{15.0}{3.8} = 3.95 \text{ min}$$

Indexing Attachment. The indexing attachment, or dividing head, is used to divide cylindrical parts, or stock, into equal divisions (spaces or angles). This operation is called *indexing*. The indexing attachment consists of the head, the tail stock, and two or three index plates (Figure 3-88).

Most indexing attachments use a worm and worm wheel with a ratio of 1:40; that is, for every 40 turns of the index crank, the spindle rotates 360°. The formulas for calculating the number of turns of the index crank for a certain number of divisions (spaces) and a certain number of degrees (angles) are:

$$T = \frac{40}{N} \quad \text{or}$$

$$T = \frac{40 \times X^\circ}{360}$$

T = turns of the index crank

N = number of desired divisions

X° = number of desired degrees

Index Plates. Index plates have concentric circles on their surfaces with equally spaced holes. These holes serve as fractions of a turn of the index crank. See Figure 3-89 for the number of holes on each standard index plate—Brown & Sharpe, and Cincinnati.

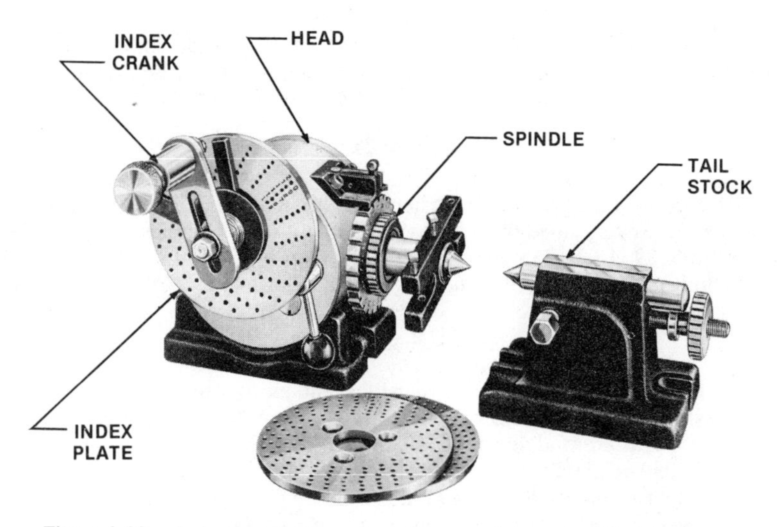

Figure 3-88. Indexing attachments and nomenclature. (L. W. Chuck Co.)

BROWN & SHARPE INDEX PLATES		CINCINNATI INDEX PLATES	
Holes on plate 1:	15 16 17 18 19 20	Holes on side 1:	24 25 28 30 34 37 38 39 41 42 43
Holes on plate 2:	21 23 27 29 31 33	Holes on side 2:	46 47 49 51 53 54 57 58 59 62 66
Holes on plate 3:	37 39 41 43 47 49	----------------	

Figure 3-89. Standard index plate holes.

When a part is to be divided into a certain number of divisions or a certain number of degrees, one of the index plates must be used. The correct index plate with the proper number of holes depends on the denominator of the fraction that results from the division in the formulas:

$$T = \frac{40}{N} \quad \text{and} \quad T = \frac{40 \times {}^\circ}{360}$$

NOTE: The number of holes in the index plate may be equal to the number of divisions desired, or to a multiple of this number.

Indexing for Equal Divisions. To divide a circumference into 49 equal divisions, an index plate with 49 holes must be used. By substituting in the formula,

$$T = \frac{40}{N}$$

it is determined that $40/49$ of a turn is required. Therefore, the index crank must be turned to cover only 40 holes. Another situation would be dividing a circumference into 13 equal divisions. By using the formula,

$$T = \frac{40}{N}$$

it is determined that $40/13 = 3\,1/13$. Therefore, an index plate with 13 holes must be used. However, since there is no index plate with 13 holes (Figure 3-89), its multiple 39 (13×3) must be used. Since an index plate with 39 holes is used, multiply the $3\,1/13$ from the original calculation, by $3/3$:

$$3\,1/13 \times 3/3 = 3\,3/39$$

Consequently, for each of the 13 divisions, the crank must be turned 3 complete revolutions plus 3 holes when using the 39-hole index plate.

These circumferential divisions (49 and 13) can be made either by using the Brown & Sharpe plate three, which has the 49 holes and 39 holes, or the Cincinnati plate two, which has the 49 holes, and the plate one, which has 39 holes. Another way to determine the number of turns required for equal divisions of a circumference is to use the tables (Figure 3-90) that provide the necessary information.

NO. OF DIV.	INDEX PLATE HOLES	TURNS	HOLES	INDEX PLATE HOLES	TURNS	HOLES	INDEX PLATE HOLES	TURNS	HOLES
1	any	40	—	any	40	—	any	40	—
2	any	20	—	any	20	—	any	20	—
3	39	13	13	33	13	11	18	13	6
4	any	10	—	any	10	—	any	10	—
5	any	8	—	any	8	—	any	8	—
6	39	6	26	33	6	22	18	6	12
7	49	5	35	21	5	15	—	—	—
8	any	5	—	any	5	—	any	5	—
9	—	—	—	27	4	12	18	4	8
10	any	4	—	any	4	—	any	4	—
11	—	—	—	33	3	21	—	—	—
12	39	3	13	33	3	11	18	3	6
13	39	3	3	—	—	—	—	—	—
14	49	2	42	21	2	18	—	—	—
15	39	2	26	33	2	22	18	2	12
16	—	—	—	—	—	—	20	2	10
17	—	—	—	—	—	—	17	2	6
18	—	—	—	27	2	6	18	2	4
19	—	—	—	—	—	—	19	2	2
20	any	2	—	any	2	—	any	2	—

Figure 3-90. Partial listing of data required for dividing circumferences into an equal number of divisions when indexing.

Indexing for Equal Angles. In addition to dividing a circumference into equal divisions or spaces, the indexing attachment can also be used to divide a circumference into equal angles. Forty revolutions of the index crank correspond to 360°. Therefore, for every turn or revolution of the index crank, the spindle rotates one-fortieth of a revolution, or $360°/40$ (9°). This relationship can be used in angular layout work. See Example 3-10.

Example 3-10: Determine the number of turns of the index crank for a layout that requires equal angles of 60°.

Given: X° = 60°, Ratio 1:40, and the Standard Brown & Sharpe index plates (Figure 4-89)

Solution: $T = \frac{40 \times X°}{360} = \frac{40 \times 60}{360} = \frac{2{,}400}{360} = 6\frac{24}{36}$ or

$$6\frac{4}{6} = 6\frac{4 \times 3}{6 \times 3} = 6\frac{12}{18}$$

For each division, turn the index crank six revolutions, plus 12 holes.

NOTE: Since a plate with 6 holes is not available, its multiple 18 (6 × 3) must be used. The plate with 18 holes is plate one (Figure 3-89).

From Example 3-10, it is determined that $6\,12/18$ revolutions of the index crank, or 6 revolutions plus 18 holes, result in the desired 60° angle. This corresponds to the relationship that one-fortieth of an index crank revolution is equal to 9° rotation of the spindle. That is,

$$6\,12/18 \text{ or } 6.666 \times 9° = 59.99° \text{ or } 60°$$

Formed Milling Cutters. Formed milling cutters belong to a special class of cutters used to duplicate interchangeable work. Gear cutting is performed with a formed milling cutter. The standard formed milling cutters used to cut gears are characterized by their *diametral pitch* (P). Diametral pitch is the ratio of the number of teeth (N) for each inch of pitch diameter (D) or (Pd) of the gear. Figure 3-91 shows the relative size of teeth produced by various diametral pitch sizes (24P - 6P). For every diametral pitch size there are eight different cutters, each of which is suitable to cut only a certain teeth range. See Figure 3-92.

Spur Gear Cutting. The spur gear (Figure 3-93) is a wheel with straight teeth that are cut parallel to the direction of rotation. The dimensions (Figure 3-94) of a spur gear are based on its diametral pitch, and for metric gears, they are based on the *module* (m). Module is the ratio of the pitch diameter to the number of teeth of the gear.

The following formulas show the relationship between diametral pitch and module:

$$P = \frac{25.4}{m} \qquad P = \text{inches}$$

$$m = \frac{25.4}{P} \qquad m = \text{millimeters (mm)}$$

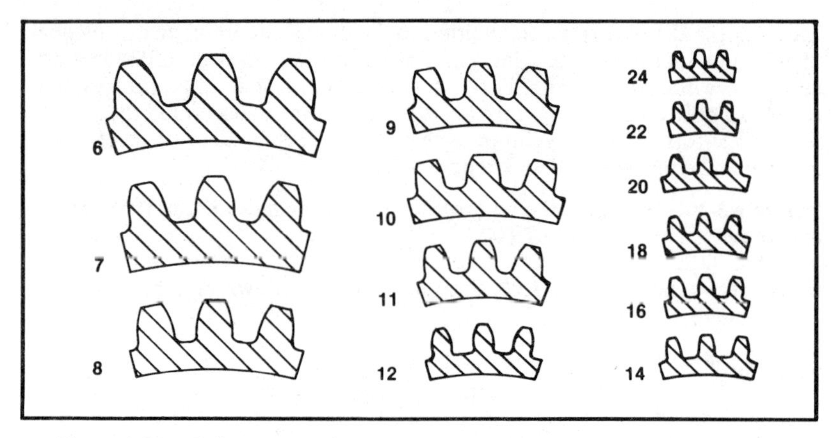

Figure 3-91. Relative size of standard gear teeth of different diametral pitch.

CUTTER NO.	1	2	3	4	5	6	7	8
TEETH RANGE	135 to rack	55 - 134	35 - 54	26 - 34	21 - 25	17 - 20	14 - 16	12 - 13

Figure 3-92. Teeth range for all standard gear cutters.

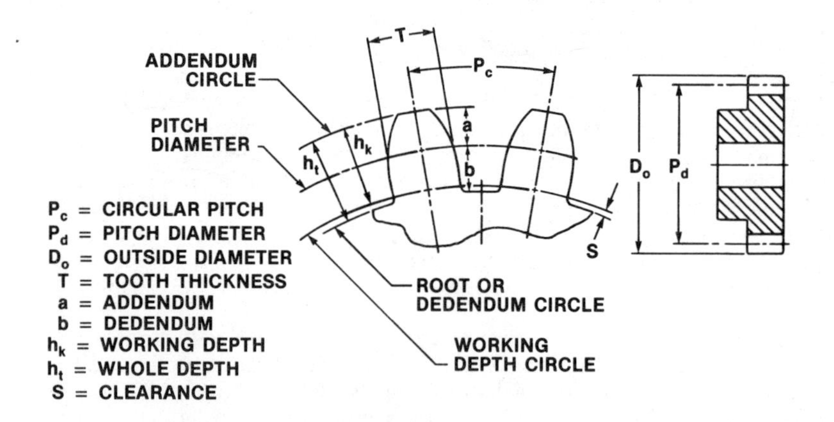

Figure 3-93. Spur gear nomenclature. (Cincinnati Milacron)

NOTE: All the dimensions of a gear are based on either the diametral pitch (English) or the module (metric).

Spur gears are easily cut on general purpose milling machines equipped with an indexing attachment. To cut a spur gear of a particular diametral pitch, the outside diameter, the whole depth-of-tooth, and the number of turns of the index crank must be known. See Example 3-11.

Example 3-11: A 34-tooth, 12-pitch spur gear is to be cut in a horizontal milling machine. Find the outside diameter (D_o), the whole depth (h_t), and the number of turns (T) of the index crank.

Given: N = 34, P = 12, Dividing head's ratio 1:40, and Standard Brown & Sharpe indexing plates

Solution: $D_o = \dfrac{N + 2}{P} = \dfrac{34 + 2}{12} = \dfrac{36}{12} = 3''$

$$h_t = \frac{2.157}{P} = \frac{2.157}{12} = 0.180''$$

$$T = \frac{40}{N} = \frac{40}{34} = 1\frac{6}{34} = 1\frac{3}{17}$$

TERM	ENGLISH UNITS		METRIC UNITS	
	Symbol	Formula	Symbol	Formula
Diametral pitch	P	$P = \dfrac{N}{P_d}$	—	—
Module	—	—	m	$m = \dfrac{D}{N}$
Circular pitch	p or P_c	$P_c = \dfrac{\pi}{P}$	P_c	$P_c = 3.1416 \times m$
Pitch diameter	D or P_d	$P_d = \dfrac{N}{P}$	D	$D = m \times N$
Outside diameter	D_o	$D_o = \dfrac{N + 2}{P}$	D_o	$D_o = m(N + 2)$ $D_o = D + 2m$
Number of teeth in the gear	N	$N = P_d \times P$	N	$N = \dfrac{D}{m}$
Tooth thickness	T	$T = \dfrac{1.5708}{P}$	T	$T = m\left(\dfrac{\pi}{2}\right)$
Addendum	a	$a = \dfrac{1}{P}$	a	$a = m$
Dedendum	b	$b = \dfrac{1.157}{P}$	b	$b = 1.25$
Working depth	h_k	$h_k = \dfrac{2}{P}$	h_k	$h_k = 2.167\ m$
Whole depth	h_t	$h_t = \dfrac{2.157}{P}$	h_t	$h_t = 2.25\ m$
Clearance	S	$S = \dfrac{0.157}{P}$	S	$S = 0.083\ m$
Center distance between meshing gears	C	$C = \dfrac{N_1 + N_2}{2P}$	C	$C = m\left(\dfrac{N_1 + N_2}{2}\right)$

Figure 3-94. Spur gear formulas.

Grinding Machine

The basic machine tool used for grinding is the grinding machine, or grinder. Many types of grinding machines, such as the cylindrical grinder (Figure 3-95) and the surface grinder (Figure 3-96), are available, each of which is designed for a specific operation.

In grinding operations, material is removed when the spindle with the grinding wheel (cutting tool) is rotating, while the table with the workpiece is providing the feed.

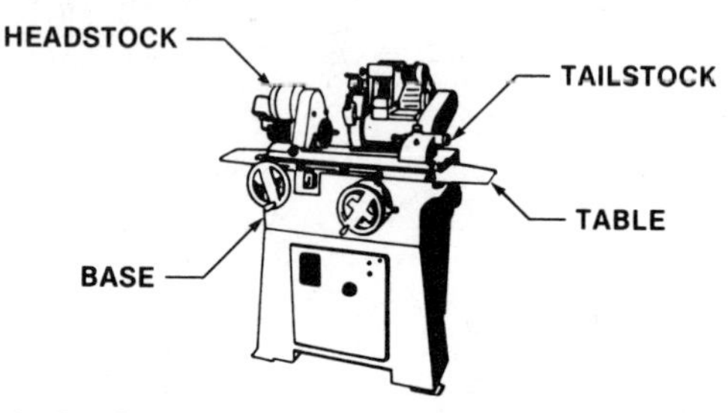

Figure 3-95. Cylindrical grinder, or grinding machine, basic nomenclature. (Clausing Division of Atlas Press Company)

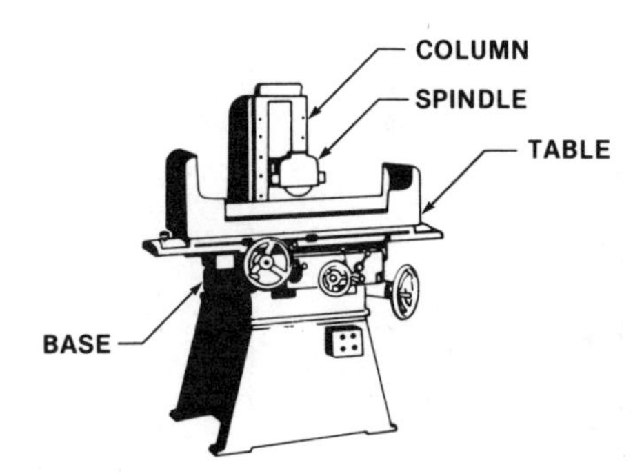

Figure 3-96. Surface grinder, or grinding machine, basic nomenclature. (Clausing Division of Atlas Press Company)

Grinding Wheel. The grinding wheel is a cutting tool used to remove material from a workpiece (stock) by a rubbing action. A grinding wheel is made of an abrasive material such as aluminum oxide, silicon carbide, or diamonds. Grinding wheels are available in a variety of faces and types (Figure 3-97).

The characteristics of a grinding wheel include:

1. Type of abrasive material.
2. Grit or grain size (number of grains of the abrasive material per linear inch).

3. Grade (degree of difficulty of separation of grains from the wheel during the grinding operation).
4. Structure (relative density of grains composing the wheel).
5. Bond type (material used in bonding or gluing the grains of the wheel).
6. Concentration (diamond only).
7. Depth of diamond section (diamond only).

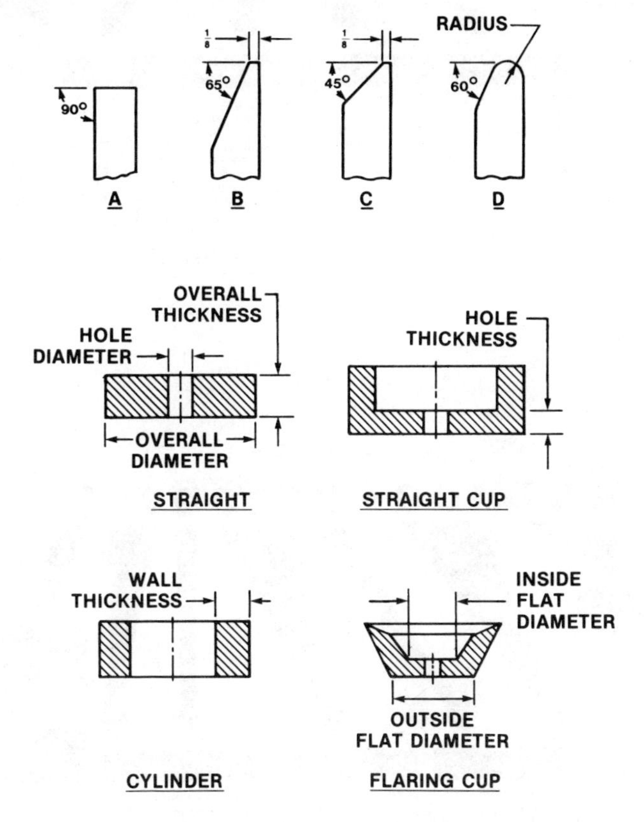

Figure 3-97. Typical standard wheel faces and types.

Standard marking systems identify the characteristics of grinding wheels (Figures 3-98 and 3-99).

The speed with which a grinding wheel is used depends upon the type of wheel and the type of machining operation involved. Figure 3-100 shows recommended grinding wheel characteristics and speeds for various grinding operations.

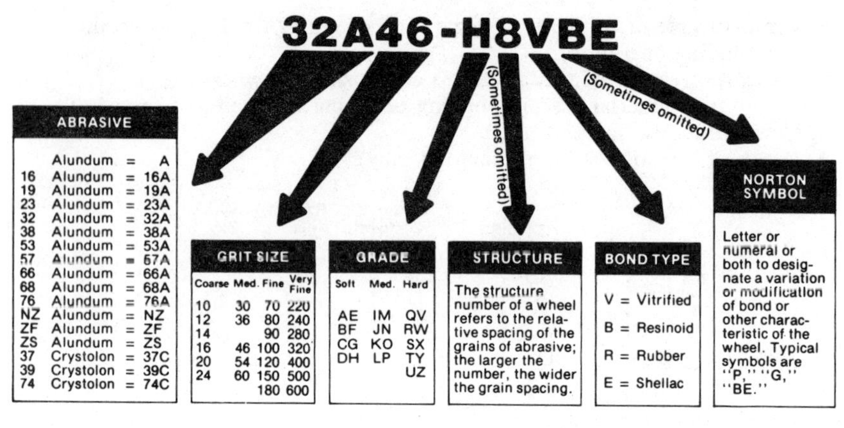

Figure 3-98. Marking system for standard grinding wheels. (Norton Company)

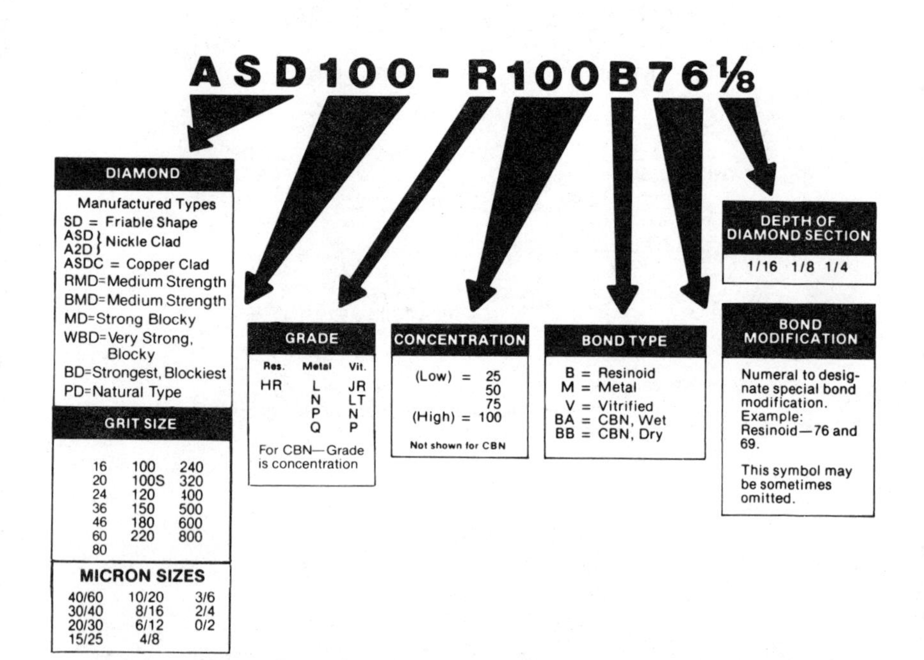

Figure 3-99. Marking system for diamond grinding wheels. (Norton Company)

WORK MATERIAL (STOCK)	GRINDING WHEEL CHARACTERISTICS			
	Abrasive Material	**Grain or Grit Size**	**Grade**	**Bond type**
Aluminum	Silicon carbide	24 - 40	K - O	Vitrified
Brass	Silicon	24 - 36	K - P	Vitrified
Bronze (hard)	Aluminum oxide	46	L	Vitrified
Cast iron	Silicon carbide	20 - 46	J - S	Vitrified
Cemented carbide (dry)	Silicon carbide	60 - 120	H - I	Vitrified
Cemented carbide (wet)	Diamond	100 - 220	R	Resinoid
Steel castings	Aluminum oxide	14 - 20	O - R	Vitrified Resinoid Rubber
Steels (hardened)	Aluminum oxide	24 - 60	I - S	Vitrified Resinoid Rubber

Application	Speed (RPM)
Cutlery (Large - Offhand)	4,000 - 5,000
Cut-off (Rubber, shellac, resinoid)	9,000 - 16,000*
Cylinders (including Hemming)	2,500 - 5,000*
Cylindrical grinding	5,000 - 8,500*
Disc grinding	4,000 - 5,500
Internal grinding	4,000 - 6,000
Knife grinding (machine knives)	3,500 - 4,500
Mounted wheels and points - Speed varies with shape, mandrel diameter, overhang (outside chuck)	
Snagging - Vitrified small hole	5,000 - 6,000
Snagging - Resinoid and rubber	7,000 - 9,500
Surface grinding	4,000 - 6,500
Tool grinding	5,000 - 6,000

*The higher speeds are recommended only where bearings, protection devices, and machine rigidity are adequate.

Figure 3-100. Recommended grinding wheel characteristics and speeds (RPM) for various grinding applications. (Norton Company)

The recommended cutting (grinding) speed (FPM) and the diameter of the grinding wheel determine the spindle speed (RPM) of any grinding machine. See Figure 3-101 for typical manufacturer spindle speed recommendations. The desired surface quality and accuracy determine the feed and depth-of-cut. The recommended feeds and depths-of-cut are:

For rough finish: Feed - 5/8 to 3/4 of the grinding wheel width per revolution of the work.
Depth-of-cut - .001″ to .0015″ or 0.025 mm to 0.038 mm.

DIAM. OF WHEEL IN INCHES	DIAM. OF WHEEL IN mm (APPROX.)	CUTTING SPEED IN FEET-PER-MINUTE (FPM) AND METERS-PER-MINUTE (m/min)					
		4,000′ 1,200 m	4,500′ 1,350 m	5,000′ 1,500 m	5,500′ 1,650 m	6,000′ 1,800 m	6,500′ 1,950 m
		SPINDLE SPEED IN REVOLUTIONS-PER-MINUTE (RPM)					
1/4	6	61,116	68,756	76,392	84,032	91,672	99,212
3/8	9	40,744	46,594	50,928	56,021	61,115	66,141
1/2	13	30,558	34,378	38,196	42,016	45,836	49,656
5/8	16	24,446	27,502	30,557	33,615	36,669	39,685
3/4	19	20,372	22,918	25,464	28,011	30,557	33,071
7/8	22	17,462	21,826	21,826	24,009	26,192	28,346
1	25	15,279	17,189	19,098	21,008	22,918	24,828
2	50	7,639	8,594	9,549	10,504	11,459	12,414
3	75	5,093	5,729	6,366	7,003	7,639	8,276
4	100	3,820	4,297	4,775	5,252	5,729	6,207
5	125	3,056	3,438	3,820	4,202	4,584	4,966
6	150	2,546	2,865	3,183	3,501	3,820	4,138
7	175	2,183	2,455	2,728	3,001	3,274	3,547
8	200	1,910	2,148	2,387	2,626	2,865	3,103
10	250	1,528	1,719	1,910	2,101	2,292	2,483

Figure 3-101. Relationship between cutting speed, spindle speed, and wheel diameter. (Norton Company)

For medium finish: Feed - 1/3 to 1/2 of the grinding wheel width per revolution of the work.
Depth-of-cut - .0002″ to .0004″ or 0.005 mm to 0.010 mm.

For fine finish: Feed - 1/4 to 1/2 of the grinding wheel width per revolution of the work.
Depth-of-cut - .0001″ to .0002″ or 0.0025 mm to 0.005 mm.

ALLOWANCES AND TOLERANCES FOR FITS

Allowances and tolerances for fits are important in the manufacturing of interchangeable parts. In machining operations, a high degree of dimensional accuracy is very expensive to achieve, and in most cases, the expense is unnecessary. By specifying allowances and tolerances, interchangeability is achieved without sacrificing quality, and at the same time, the cost of production is significantly reduced.

Every fit (the tightness of two mating parts) is characterized by the following:

1. Nominal size - the basic dimension of a part.
2. Tolerance - the permissible variation (error) in the dimension of a part from its nominal size.
3. Dimensional limits - the resulting maximum and minimum dimensions of toleranced parts.
4. Allowance - the intentional clearance or interference between two mating parts.
5. Limits of clearance - the resulting maximum and minimum allowance from the limits (upper and lower) for each class of fit.

Tolerance Position and Dimensional Limits of Parts

Tolerance position (Figure 3-102) is the relative position of a specified tolerance to the nominal size. It can either be *unilateral* or *bilateral*. The tolerance position is unilateral when the total amount is assigned in one direction, either plus (+) or minus (−). It is bilateral when the total amount is divided equally or unequally; that is, one part is assigned in one direction (+) and the other part is assigned in the other direction (−). For example, if the nominal size is 2.000″ and the tolerance is 0.004″, the unilateral tolerance position is

$$2.000 \begin{smallmatrix} +\ 0.004 \\ -\ 0.000 \end{smallmatrix} \text{ or } 2.000 \begin{smallmatrix} +\ 0.000 \\ -\ 0.004 \end{smallmatrix}$$

The bilateral tolerance position would be

$$2.000 \begin{smallmatrix} +\ 0.002 \\ -\ 0.002 \end{smallmatrix} \text{ or } 2.000 \begin{smallmatrix} +\ 0.001 \\ -\ 0.003 \end{smallmatrix}$$

$$T_s = S_M - S_m$$

$$T_h = H_M - H_m$$

$$A_m = H_m - S_M$$

$$A_M = H_M - S_m$$

TOLERANCE POSITION

S_m = MINIMUM DIAMETER OF SHAFT (LOWER LIMIT)
S_M = MAXIMUM DIAMETER OF SHAFT (UPPER LIMIT)
H_m = MINIMUM DIAMETER OF HOLE (LOWER LIMIT)
H_M = MAXIMUM DIAMETER OF HOLE (UPPER LIMIT)
T_s = TOLERANCE OF SHAFT
T_h = TOLERANCE OF HOLE
A_m = MINIMUM ALLOWANCE (LOWER LIMIT OF CLEARANCE)
A_M = MAXIMUM ALLOWANCE (UPPER LIMIT OF CLEARANCE)

Figure 3-102. A graphic explanation of fit and its components.

Specification of Tolerance

One of the most difficult problems in mass production is the specification of proper tolerances to achieve interchangeability of parts at the lowest possible cost, and still maintain quality products. If the functioning of a part is not affected by a large tolerance, the specification of a small tolerance would unnecessarily increase the cost of production. Tolerances are specified according to the size of the part, the function of the part, and the relative speed of movement between mating parts.

In general, four ranges of tolerances for parts exist:

1. 1/64″ to 1/16″ - for heavy machinery and large stationary structures.
2. 0.001″ to 0.005″ - for general classes of rotating machinery, including many parts of automobiles and trucks.

3. 0.0001″ to 0.0005″ - for high-speed machinery such as aircraft engine parts and critical parts of automobile engines.
4. 0.00001″ to 0.0001″ - for precision instruments, precise ball and roller bearings, and any components requiring precision.

Precision Capability of Machine Tools

The degree of accuracy of a machined part depends upon the particular machining process and the condition of the machine tool. Every machine tool has a minimum capability for precision that it cannot exceed. Therefore, when specifying tolerances, the capability of the machine tool must be considered. For example, as Figure 3-103 shows, the smallest tolerance obtained in drilling operations is between 0.002″ and 0.010″.

SIZE RANGE		TOLERANCES (TOTAL)								
FROM	TO AND INCLUDED									
.000	.599	.00015	.0002	.0003	.0005	.0008	.0012	.002	.003	.005
.600	.999	.00015	.00025	.0004	.0006	.001	.0015	.0025	.004	.007
1.000	1.499	.0002	.0003	.0005	.0008	.0012	.002	.003	.005	.008
1.500	2.799	.00025	.0004	.0006	.001	.0015	.002	.004	.006	.010
2.800	4.499	.0003	.0005	.0008	.0012	.002	.003	.005	.008	.012
4.500	7.799	.0004	.0006	.001	.0015	.0025	.004	.006	.010	.015
7.800	15.599	.0005	.0008	.0012	.002	.003	.005	.008	.012	.019
15.600	20.999	.0006	.001	.0015	.0025	.004	.006	.010	.015	.025

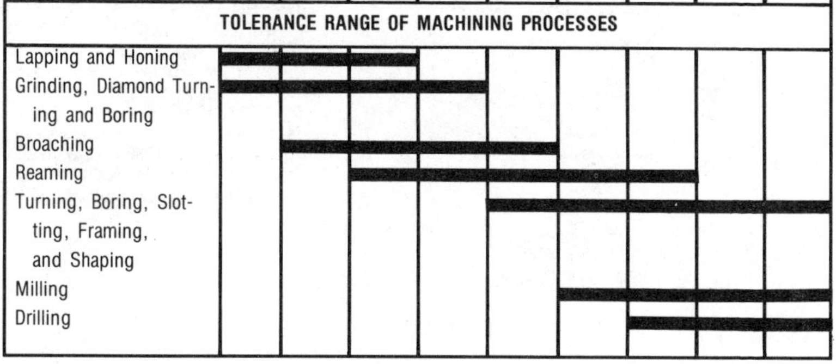

TOLERANCE RANGE OF MACHINING PROCESSES

Lapping and Honing
Grinding, Diamond Turning and Boring
Broaching
Reaming
Turning, Boring, Slotting, Framing, and Shaping
Milling
Drilling

Figure 3-103. Tolerance selection guide.

Standard ANSI Fits

The American National Standards Institute (ANSI) has standardized tolerances and allowances for various classes of five types of fits:

1. *Running or sliding fits* (RC) allow relative freedom between the mating parts.
2. *Locational clearance fits* (LC) allow mating parts to be assembled and disassembled freely.
3. *Transition fits or Interference fits* (LT) allow small amounts of either clearance, or interference, of the mating parts.
4. *Locational interference fits* (LN) allow a certain amount of interference so that the mating parts can be assembled only by force.

5. *Force of Shrink fits* (FN) allow larger amounts of interference. The mating parts are assembled either by applying great pressure on the part, or by heating the bore then forcing the shaft inside it.

The symbols for type of fit is followed by a number indicating its class. Figure 3-104 shows typical classes of running or sliding fits. The minimum and maximum clearance (limits of clearance) of any fit depend upon the amount of tolerance and the tolerance position of both the hole and the shaft. See Example 3-12.

Example 3-12: Determine the dimensional limits of the mating parts of a class RC2 fit that has a nominal size of 0.50″. (Refer to Figures 3-102 and 3-104.)

Given: Hole tolerance T_H: $^{+0.4}_{-0.0}$ Shaft tolerance T_S: $^{-0.25}_{-0.55}$

Solution: Dimensional limits of hole: $0.500\ ^{+0.4}_{-0.0} = H_m = 0.500''$ and $H_M = 0.5004''$
Dimensional limits of shaft: $0.500\ ^{-0.25}_{-0.55} = S_m = 0.49945''$ and $S_M = 0.49975''$

Nominal Size Range Inches		Class RC 1			Class RC 2		
		Limits of Clearance	Standard Limits		Limits of Clearance	Standard Limits	
Over	To		Hole H5	Shaft g4		Hole H6	Shaft g5
0	– 0.12	**0.1** **0.45**	+ **0.2** **0**	– **0.1** – **0.25**	**0.1** **0.55**	+ **0.25** **0**	– **0.1** – **0.3**
0.12	– 0.24	**0.15** **0.5**	+ **0.2** **0**	– **0.15** – **0.3**	**0.15** **0.65**	+ **0.3** **0**	– **0.15** – **0.35**
0.24	– 0.40	**0.2** **0.6**	**0.25** **0**	– **0.2** – **0.35**	**0.2** **0.85**	+ **0.4** **0**	– **0.2** – **0.45**
0.40	– 0.71	**0.25** **0.75**	+ **0.3** **0**	– **0.25** – **0.45**	**0.25** **0.95**	+ **0.4** **0**	– **0.25** – **0.55**

NOTE: LIMITS ARE IN THOUSANDTHS OF AN INCH.

Figure 3-104. Typical data for selected running and sliding fits. (The American Society of Mechanical Engineers)

Standard ISO Tolerance

The International Organization for Standardization (ISO) has standardized 18 tolerance grades. Each of these grades provides a certain amount of tolerance and is indicated by a number. Figure 3-105 shows a partial listing of these tolerance grades.

NOMINAL SIZE IN mm	TOLERANCE GRADES								
	4	5	6	7	8	9	10	11	12
1 - 3	3	4	6	10	14	25	40	60	100
3 - 6	4	5	8	12	18	30	48	75	120
6 - 10	4	6	9	15	22	36	58	90	150
10 - 18	5	8	11	18	27	43	70	110	180
18 - 30	6	9	13	21	33	52	84	130	210
30 - 50	7	11	16	25	39	62	100	160	250
50 - 80	8	13	19	30	46	74	120	190	300
80 - 120	10	15	22	35	54	87	140	220	350
120 - 180	12	18	25	40	63	100	160	250	400
180 - 250	14	20	29	46	72	115	185	290	460
250 - 315	16	23	32	52	81	130	210	320	520
315 - 400	18	25	36	57	89	140	230	360	570
400 - 500	20	27	40	63	97	155	250	400	630

NOTE: AMOUNT OF TOLERANCE IN μm (0.001 mm).

Figure 3-105. Standard ISO tolerances for selected grades (4–12).

Symbols for ISO Tolerances

In addition to the amount of tolerance indicated by a tolerance grade number, the ISO standards also specify the tolerance position by using letters. Uppercase letters indicate the tolerance position of the hole while lower case letters indicate tolerance position of the shaft. Figure 3-106 shows the tolerance position of each letter in relation to the basic size.

The *centerline*, or *baseline* in Figure 3-106, is identified with zero (0). It represents the nominal size, or basic dimension, of a part in a drawing. This nominal size is used to specify tolerances. Specified tolerances with a plus sign are represented by letters above the baseline. Specified tolerances with a minus sign are represented by letters below the baseline.

Each letter has a definite relationship to the baseline. For example, the uppercase letter H is located above the line and the lowercase letter h is located below the baseline, but both letters touch the baseline. Therefore, the total amount of tolerance is placed unilaterally, either as $^{+\text{ tolerance}}_{-\text{ zero}}$ or $^{+\text{ zero}}_{-\text{ tolerance}}$. The tolerance of the uppercase letter H always has a plus sign and the lowercase letter h is always a minus sign.

The amount of tolerance and its relative position to the baseline for tolerance grades 4 to 12, and letters H and h can be found by referring to Figures 3-105 and 3-106. For example, if given a nominal size of 40 mm, a tolerance grade of 9, and the letters H and h, the amount of tolerance is 62 μm or 0.062 mm (Figure 3-105). The position of this tolerance is plus for H and minus for h; that is,

$$40\ H_9 = 40\ ^{+\ 62}_{-\ 0} \text{ and } 40\ h_9 = 40\ ^{+\ 0}_{-\ 62}$$

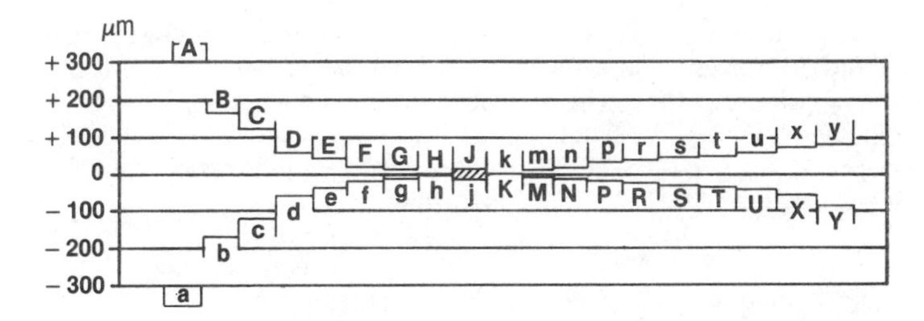

Figure 3-106. Relative tolerance position for each letter symbol.

Standard ISO Fits

Standard ISO fits are divided into three categories:

1. *Clearance fits* - include loose (coarse) fits, medium fits, and close (fine) fits. The mating parts can be assembled freely.
2. *Transition fits* - include locational transition fits in which the mating parts can be assembled only selectively because the clearance is approximately zero between the mating parts.
3. *Interference fits* - include locational interference fits, medium drive fits, and force fits. The mating parts can be assembled only with pressure or force. Interference fits allow negative clearance (interference).

NOTE: Tolerance grade and tolerance position are specified for each of the mating parts, therefore, the amount of clearance or interference of a particular fit depends upon the tolerance grade and tolerance position of both the hole and the shaft.

Below is a representation of a typical standard ISO fit:

$$\frac{45H_7}{45h_6} \quad \text{or} \quad 45\,\frac{H_7}{h_6}$$

45: nominal size of the fit (45 mm) common to both the hole and shaft.
H_7: tolerance position and tolerance grade of the hole.
h_6: tolerance position and tolerance grade of the shaft.

The amount of tolerance and the tolerance position of the hole and the shaft according to the ISO Standards for tolerances are:

$$\text{Hole: } 45H_7 = 45\,^{+25}_{-0} = 45.000 \text{ - } 45.025 \text{ mm}$$

$$\text{Shaft: } 45h_6 = 45\,^{+0}_{-16} = 44.984 \text{ mm - } 45.000 \text{ mm}$$

NOTE: The amount of tolerance corresponding to the nominal size between 30 mm and 50 mm is 16 μm (0.016 mm) for grade 6, and 25 μm (0.25 mm) for grade 7 (Figure 3-105). The tolerance position for letter H is positive, or above the baseline, and for letter h, the tolerance position is negative, or below the baseline (Figure 3-106). (Refer to Standard ISO Tolerances and Symbols for ISO Tolerances and additional information.)

INSPECTION IN MANUFACTURING

Various tools, or devices, are used to inspect manufactured parts. Among the common devices are the *gauge block, sine bar,* and *fixed-limit gauge.*

Gauge Block

Gauge blocks (Figure 3-107) are used as standards of length. They are rectangular and have two flat and parallel surfaces. They are available in sets (Figure 3-108).

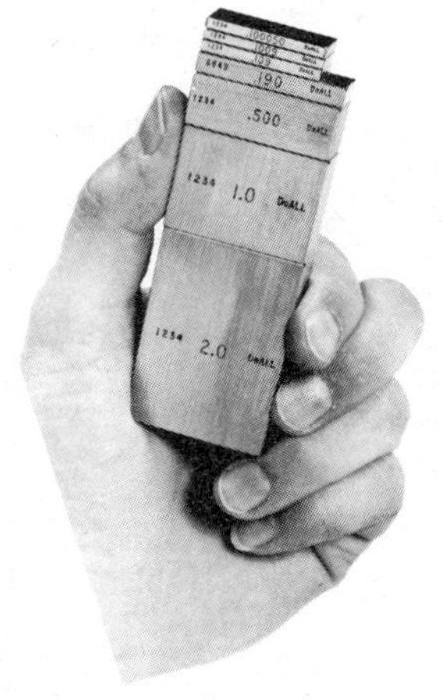

Figure 3-107. Gauge blocks. (DoAll Company)

The number of gauge blocks in a set range from 5 to 120. The quality of gauge blocks depends upon the grade. Four grades are available and each of these grades is suitable for certain uses:

1. Grade AAA Blocks have the least amount of tolerance and are used as reference standards for the highest precision possible.
2. Grade AA Blocks are used to check and calibrate other grades of gauge blocks.
3. Grade A+ or A Blocks are precision grade blocks and are used for quality cost inspection.
4. Grade B Blocks have greater tolerance than the other gauge blocks and are used in the workshop during production.

Figure 3-108. Gauge block set. (DoAll Company)

The tolerances specified for all four grades of gauge blocks are very small. Using several blocks to build a reference dimension affects the overall, or total length little because each grade block is accurate to within .000008″.

Gauge Block Sets. Two of the most popular sets of gauge blocks are the English-unit set of 81 blocks, and the metric-unit set of 111 blocks (Figure 3-109).

The gauge blocks from the English set are used to build up standard reference lengths in inches. The gauge blocks from the metric set are used to build up lengths in millimeters.

ENGLISH SET OF 81 BLOCKS			METRIC SET OF 111 BLOCKS		
Pieces	Series	Range of Length	Pieces	Series	Range of Length
9	.0001″	.1001 - .1009″	9	0.001 mm	1.001 - 1.009 mm
49	.001″	.101 - .149″	49	0.01 mm	1.01 - 1.49 mm
19	.050″	.050 - .950″	49	0.5 mm	0.5 - 25.5 mm
4	1.000″	1.000 - 4.000″	4	25 mm	25 - 100 mm

Figure 3-109. Dimensions of two frequently used gauge block sets.

When selecting gauge blocks from a set to build up a standard reference length, begin with the block that will eliminate the number in the fourth decimal place, then the third, and so forth. See Example 3-13 for an application of this operation.

Example 3-13: Find the gauge blocks needed to build up the standard lengths 1.8546″ and 79.328 mm.

Solution:

English Gauge Blocks	
Total length:	1.8546
First block:	−.1006
	1.7540
Second block:	−.104
	1.650
Third block:	−.650
	1.000
Fourth block:	−1.000
	0.000

Metric Gauge Blocks	
Total length:	79.328
First block:	−1.008
	78.320
Second block:	−3.32
	75.00
Third block:	−75.00
	00.00

Sine Bar

The sine bar (Figure 3-110) is a device consisting of a hardened steel plate with two cylindrical pins of the same diameter, attached at its ends. The sine bar is used with gauge blocks and a dial indicator for the inspection of tapers (Figure 3-111).

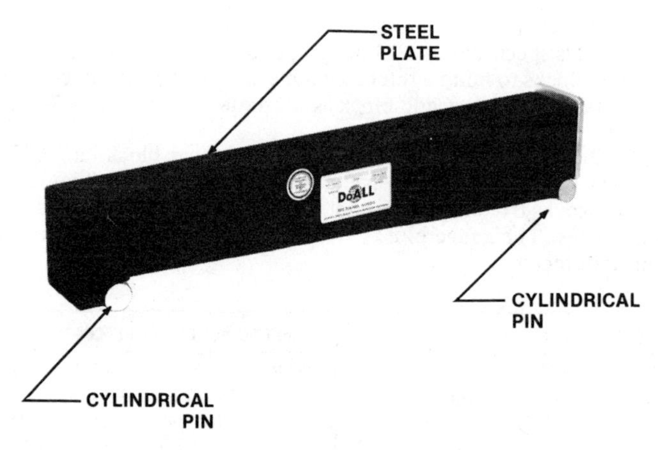

Figure 3-110. Sine bar. (DoAll Company)

In any sine bar setting, there is a trigonometric problem to be solved that involves a right triangle with two known factors. The first of these factors is the hypotenuse of the triangle represented by the length (L) of the sine bar. Depending on the sine bar used, the length is equal to either 5″ or 10″. The second factor is the side opposite, represented by the height (h). The height is equal to the total length of the gauge blocks used to build the height for a particular setting. See Examples 3-14 and 3-15.

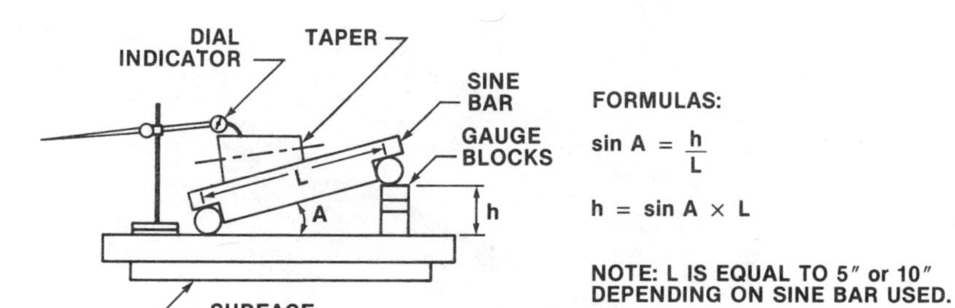

Figure 3-111. Sine bar setting and formulas.

Example 3-14: A tapered plug with an included angle of 20° is to be inspected with a 5″ sine bar. Find the height (h) and select the appropriate gauge blocks to establish this height.

Given: $A = 20°$ $\sin 20° = 0.34202$, $L = 5''$

Solution: $h = \sin A \times L = 0.34202 \times 5 = 1.7101''$

To establish this length, use the following gauge blocks:
$0.1001 + 0.110 + 0.500 + 1.000 = 1.7101''$

Example 3-15: The total length of the gauge blocks used in a 10″ sine bar setting is 2.6842″. Find the included angle (A) that corresponds to this setting.

Given: $L = 10''$, $h = 2.6842''$

Solution: $\sin A = \dfrac{h}{L} = \dfrac{2.6842}{10} = 0.26842$

$A = 15°\,33'$

Fixed-limit Gauge

Fixed-limit gauges (Figures 3-112 and 3-113) are standard gauges used for the inspection of toleranced parts. They are characterized by the two dimensions, "GO" and "NOT GO". The "GO" dimension represents the lower dimen-

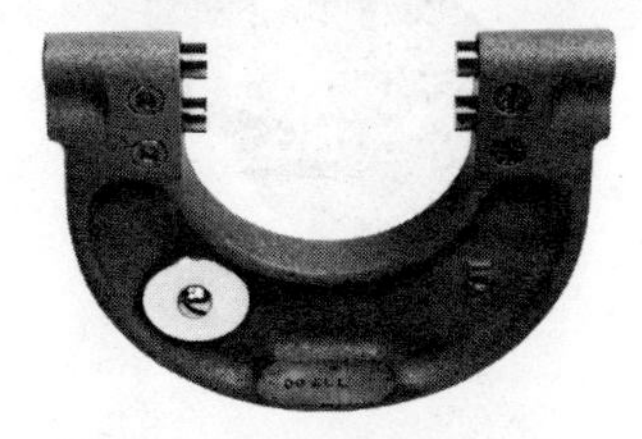

Figure 3-112. Fixed-limit gauge for external cylindrical inspection. (DoAll Company)

Figure 3-113. Fixed-limit gauge for internal cylindrical inspection. (DoAll Company)

sional limit and the "NOT GO" represents the upper dimensional limit in accordance with the specified tolerance for a certain nominal size. The fixed-limit gauges shown in Figures 3-112 and 3-113 are used to inspect cylindrical parts.

Fixed-limit screw thread gauges (Figures 3-114 and 3-115) are used to inspect standard screw threads. The dimensional limits of toleranced screw threads of various classes and fits (English and metric threads) are shown in Figures 3-116, 3-117, and 3-118.

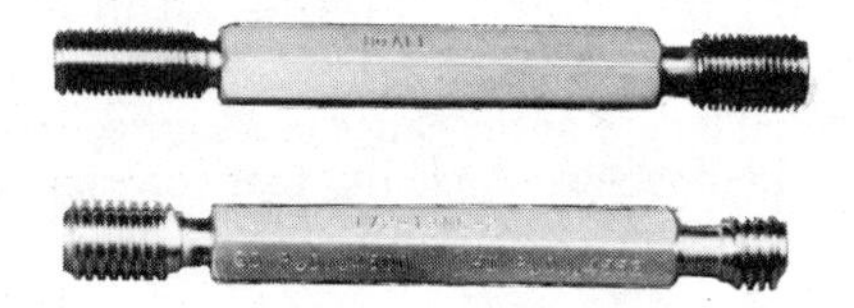

Figure 3-114. Fixed-limit screw thread gauge for internal thread inspection. (DoAll Company)

Figure 3-115. Fixed-limit screw thread gauge for external thread inspection. (DoAll Company)

NOMINAL SIZE	AMERICAN NATIONAL			UNIFIED NATIONAL			
	GO	NO GO		GO		NO GO	
	CL 2 & 3	CL 2	CL 3	CL 3A	CL 2A	CL 2A	CL 3A
#0-80 NF	.0519	.0502	.0506	.0519	.0514	.0496	.0506
#1-64 NC	.0629	.0610	.0615	.0629	.0623	.0603	.0614
#1-72 NF	.0640	.0622	.0627	.0640	.0634	.0615	.0626
#2-56 NC	.0744	.0724	.0729	.0744	.0738	.0717	.0728
#2-64 NF	.0759	.0740	.0745	.0759	.0753	.0733	.0744
#3-48 NC	.0855	.0833	.0839	.0855	.0848	.0825	.0838
#3-56 NF	.0874	.0854	.0859	.0874	.0867	.0845	.0858
#4-40 NC	.0958	.0934	.0941	.0958	.0950	.0925	.0939
#4-48 NF	.0985	.0963	.0969	.0985	.0978	.0954	.0967
#5-40 NC	.1088	.1064	.1071	.1088	.1080	.1054	.1069
#5-44 NF	.1102	.1079	.1086	.1102	.1095	.1070	.1083
#6-32 NC	.1177	.1150	.1158	.1177	.1169	.1141	.1156
#6-40 NF	.1218	.1194	.1201	.1218	.1210	.1184	.1198
#8-32 NC	.1437	.1410	.1418	.1437	.1428	.1399	.1415
#8-36 NF	.1460	.1435	.1442	.1460	.1452	.1424	.1439
#10-24 NC	.1629	.1596	.1605	.1629	.1619	.1586	.1604
#10-32 NF	.1697	.1670	.1678	.1697	.1688	.1658	.1674
#12-24 NC	.1889	.1856	.1865	.1889	.1879	.1845	.1863
#12-28 NF	.1928	.1897	.1906	.1928	.1918	.1886	.1904
#12-32 NEF	.1957	.1926	.1935	.1957	.1948	.1917	.1933
¼-20 NC, UNC	.2175	.2139	.2149	.2175	.2164	.2127	.2147
¼-28 NF, UNF	.2268	.2237	.2246	.2268	.2258	.2225	.2243
¼-32 NEF	.2297	.2265	.2275	.2297	.2287	.2255	.2273

NOMINAL SIZE	GO BASIC ALL CLASSES ALL SERIES	AMERICAN NATIONAL		UNIFIED NATIONAL	
		GO	NO GO		GO
		CL 2	CL 3	CL 2B	CL 3B
#0-80 NF	.0519	.0536	.0532	.0542	.0536
#1-64 NC	.0629	.0648	.0643	.0655	.0648
#1-72 NF	.0640	.0658	.0653	.0665	.0659
#2-56 NC	.0744	.0764	.0759	.0772	.0765
#2-64 NF	.0759	.0778	.0773	.0786	.0779
#3-48 NC	.0855	.0877	.0871	.0885	.0877
#3-56 NF	.0874	.0894	.0889	.0902	.0895
#4-40 NC	.0958	.0982	.0975	.0991	.0982
#4-48 NF	.0985	.1007	.1001	.1016	.1008
#5-40 NC	.1088	.1112	.1105	.1121	.1113
#5-44 NF	.1102	.1125	.1118	.1134	.1126
#6-32 NC	.1177	.1204	.1196	.1214	.1204
#6-40 NF	.1218	.1242	.1235	.1252	.1243
#8-32 NC	.1437	.1464	.1456	.1475	.1465
#8-36 NF	.1460	.1485	.1478	.1496	.1487
#10-24 NC	.1629	.1662	.1653	.1672	.1661
#10-32 NF	.1697	.1724	.1716	.1736	.1726
#12-24 NC	.1889	.1922	.1913	.1933	.1922
#12-28 NF	.1928	.1959	.1950	.1970	.1959
#12-32 NEF	.1957	.1988	.1979	.1998	.1988
¼-20 NC, UNC	.2175	.2211	.2201	.2224	.2211
¼-28 NF, UNF	.2268	.2299	.2290	.2311	.2300
¼-32 NEF	.2297	.2329	.2319	.2339	.2328

continued

NOMINAL SIZE	AMERICAN NATIONAL			UNIFIED NATIONAL			
	GO	NO GO		GO		NO GO	
	CL 2 & 3	CL 2	CL 3	CL 3A	CL 2A	CL 2A	CL 3A
5/16-18 NC, UNC	.2764	.2723	.2734	.2764	.2752	.2712	.2734
5/16-24 NF, UNF	.2854	.2821	.2830	.2854	.2843	.2806	.2827
5/16-32 NEF	.2922	.2889	.2899	.2922	.2912	.2880	.2898
3/8-16 NC, UNC	.3344	.3299	.3312	.3344	.3331	.3287	.3311
3/8-24 NF, UNF	.3479	.3446	.3455	.3479	.3468	.3430	.3450
3/8-32 NEF	.3547	.3513	.3523	.3547	.3537	.3503	.3522
7/16-14 NC, UNC	.3911	.3862	.3875	.3911	.3897	.3850	.3876
7/16-20 NF, UNF	.4050	.4014	.4024	.4050	.4037	.3995	.4019
7/16-28 NEF, UNEF	.4143	.4107	.4118	.4143	.4132	.4096	.4116
1/2-12 N	.4459	.4403	.4419	.4459	.4443	.4389	.4419
1/2-13 NC, UNC	.4500	.4448	.4463	.4500	.4485	.4435	.4463
1/2-20 NF, UNF	.4675	.4639	.4649	.4675	.4662	.4619	.4643
1/2-28 NEF, UNEF	.4768	.4731	.4742	.4768	.4757	.4720	.4740
9/16-12 NC, UNC	.5084	.5028	.5044	.5084	.5068	.5016	.5045
9/16-18 NF, UNF	.5264	.5223	.5234	.5264	.5250	.5205	.5230
9/16-24 NEF	.5354	.5314	.5326	.5354	.5342	.5303	.5325
5/8-11 NC, UNC	.5660	.5601	.5618	.5660	.5644	.5589	.5619
5/8-18 NF, UNF	.5889	.5848	.5859	.5889	.5875	.5828	.5854
5/8-24 NEF	.5979	.5938	.5950	.5979	.5967	.5927	.5949
11/16-24 NEF	.6604	.6563	.6575	.6604	.6592	.6552	.6574
3/4-10 NC, UNC	.6850	.6786	.6805	.6850	.6832	.6773	.6806
3/4-16 NF, UNF	.7094	.7049	.7062	.7094	.7079	.7029	.7056
3/4-20 NEF, UNEF	.7175	.7129	.7143	.7175	.7162	.7118	.7142

NOMINAL SIZE	GO BASIC ALL CLASSES ALL SERIES	AMERICAN NATIONAL		UNIFIED NATIONAL	
		GO	NO GO		GO
		CL 2	CL 3	CL 2B	CL 3B
5/16-18 NC, UNC	.2764	.2805	.2794	.2817	.2803
5/16-24 NF, UNF	.2854	.2887	.2878	.2902	.2890
5/16-32 NEF	.2922	.2955	.2945	.2964	.2953
3/8-16 NC, UNC	.3344	.3389	.3376	.3401	.3387
3/8-24 NF, UNF	.3479	.3512	.3503	.3528	.3516
3/8-32 NEF	.3547	.3581	.3571	.3591	.3580
7/16-14 NC, UNC	.3911	.3960	.3947	.3972	.3957
7/16-20 NF, UNF	.4050	.4086	.4076	.4104	.4091
7/16-28 NEF, UNEF	.4143	.4179	.4168	.4189	.4178
1/2-12 N	.4459	.4515	.4499	.4529	.4511
1/2-13 NC, UNC	.4500	.4552	.4537	.4565	.4548
1/2-20 NF, UNF	.4675	.4711	.4701	.4731	.4717
1/2-28 NEF, UNEF	.4768	.4805	.4794	.4816	.4804
9/16-12 NC, UNC	.5084	.5140	.5124	.5152	.5135
9/16-18 NF, UNF	.5264	.5305	.5294	.5323	.5308
9/16-24 NEF	.5354	.5394	.5382	.5405	.5392
5/8-11 NC, UNC	.5660	.5719	.5702	.5732	.5714
5/8-18 NF, UNF	.5889	.5930	.5919	.5949	.5934
5/8-24 NEF	.5979	.6020	.6008	.6031	.6018
11/16-24 NEF	.6604	.6645	.6633	.6656	.6643
3/4-10 NC, UNC	.6850	.6914	.6895	.6927	.6907
3/4-16 NF, UNF	.7094	.7139	.7126	.7159	.7143
3/4-20 NEF, UNEF	.7175	.7221	.7207	.7232	.7218

Figure 3-116. Partial listing of pitch diameters for various classes of English threads. (DoAll Company)

NOM. DIA.	PITCH	CLOSE (FINE) FIT = 4H5H					MEDIUM FIT = 6H					LOOSE (COARSE) FIT = 7H				
		MAJOR DIA.	PITCH DIAMETER		MINOR DIAMETER		MAJOR DIA.	PITCH DIAMETER		MINOR DIAMETER		MAJOR DIA.	PITCH DIAMETER		MINOR DIAMETER	
		MIN. (must clear)	MIN.	MAX.	MIN.	MAX.	MIN. (must clear)	MIN.	MAX.	MIN.	MAX.	MIN. (must clear)	MIN.	MAX.	MIN.	MAX.
3.0	0.5	3.0	2.675	2.738	2.459	2.571	3.0	2.675	2.775	2.459	2.599	3.0	2.800	2.675	2.459	2.639
3.5	0.6	3.5	3.110	3.181	2.850	2.975	3.5	3.110	3.222	2.850	3.010	3.5	3.250	3.110	2.850	3.050
4.0	0.7	4.0	3.545	3.620	3.242	3.382	4.0	3.545	3.663	3.242	3.422	4.0	3.695	3.545	3.242	3.466
4.5	0.75	4.5	4.013	4.088	3.688	3.838	4.5	4.013	4.131	3.688	3.878	4.5	4.163	4.013	3.688	3.924
5.0	0.8	5.0	4.480	4.580	4.134	4.294	5.0	4.480	4.605	4.134	4.334	5.0	4.640	4.480	4.134	4.384
6.0	1.0	6.0	5.350	5.445	4.917	5.107	6.0	5.350	5.500	4.917	5.153	6.0	5.540	5.350	4.917	5.217
7.0	1.0	7.0	6.350	6.445	5.917	6.107	7.0	6.350	6.500	5.917	6.153	7.0	6.540	6.350	5.917	6.217
8.0	1.25	8.0	7.188	7.288	6.647	6.859	8.0	7.188	7.348	6.647	6.912	8.0	7.388	7.188	6.647	6.982
9.0	1.25	9.0	8.188	8.288	7.647	7.859	9.0	8.188	8.348	7.647	7.912	9.0	8.388	8.188	7.647	7.982
10.0	1.5	10.0	9.026	9.138	8.376	8.612	10.0	9.026	9.206	8.376	8.676	10.0	9.250	9.026	8.376	8.751
11.0	1.5	11.0	10.026	10.138	9.376	9.612	11.0	10.026	10.206	9.376	9.676	11.0	10.250	10.026	9.376	9.751
12.0	1.75	12.0	10.863	10.988	10.106	10.371	12.0	10.863	11.063	10.106	10.441	12.0	11.113	10.863	10.106	10.531

Figure 3-117. Partial listing of major pitch/minor diameters for various fits of standard external metric threads. (DoAll Company)

NOM. DIA.	PITCH	CLOSE (FINE) FIT = 4h						MEDIUM FIT = 6g						LOOSE (COARSE) FIT = 8g					
		MAJOR DIAMETER		PITCH DIAMETER		MINOR DIAMETER		MAJOR DIAMETER		PITCH DIAMETER		MINOR DIAMETER		MAJOR DIAMETER		PITCH DIAMETER		MINOR DIAMETER	
		MAX.	MIN.	MAX.	MIN.	MAX.	MIN.	MAX.	MIN.	MAX.	MIN.	MAX.	MIN.	MAX.	MIN.	MAX.	MIN.	MAX.	MIN.
3.0	0.5	3.0	2.933	2.675	2.627	2.387	2.303	2.980	2.874	2.655	2.580	2.367	2.256	—	—	—	—	—	—
3.5	0.6	3.0	3.420	3.110	3.057	2.764	2.668	3.479	3.354	3.089	3.004	2.743	2.615	—	—	—	—	—	—
4.0	0.7	4.0	3.910	3.545	3.489	3.141	3.035	3.978	3.838	3.523	3.438	3.119	2.979	—	—	—	—	—	—
4.5	0.75	4.0	4.410	4.013	3.957	3.580	3.470	4.478	4.338	3.991	3.901	3.558	3.414	—	—	—	—	—	—
5.0	0.8	5.0	4.905	4.480	4.420	4.019	3.901	4.976	4.826	4.456	4.361	3.995	3.842	4.976	4.740	4.456	4.306	3.995	3.787
6.0	1.0	6.0	5.888	5.350	5.279	4.773	4.630	5.974	5.794	5.324	5.212	4.747	4.563	5.974	5.694	5.324	5.144	4.747	4.495
7.0	1.0	7.0	6.888	6.350	6.279	5.773	5.630	6.974	6.794	6.324	6.212	5.747	5.563	6.974	6.694	6.324	6.144	5.747	5.495
8.0	1.25	8.0	7.868	7.188	7.113	6.466	6.301	7.972	7.760	7.160	7.042	6.438	6.230	7.972	7.637	7.160	6.970	6.438	6.158
9.0	1.25	9.0	8.868	8.188	8.113	7.466	7.301	8.972	8.760	8.160	8.042	7.438	7.230	8.972	8.637	8.160	7.970	7.438	7.158
10.0	1.5	10.0	9.850	9.026	8.941	8.160	7.967	9.968	9.732	8.994	8.862	8.128	7.888	9.968	9.593	8.994	8.782	8.128	7.808
11.0	1.5	11.0	10.850	10.026	9.941	9.160	8.967	10.968	10.732	9.994	9.862	9.128	8.888	10.968	10.593	9.994	9.782	9.128	8.808
12.0	1.75	12.0	11.830	10.863	10.768	9.853	9.632	11.966	11.701	10.829	10.679	9.819	9.543	11.966	11.541	10.829	10.593	9.819	9.457

Figure 3-118. Partial listing of major pitch/minor diameters for various fits of standard internal metric threads. (DoAll Company)

Chapter 4

MECHANICS

Mechanics is a branch of Physical Science dealing with the study of energy and force, and their relation to the equilibrium, or motion of solids, liquids, and gases. It addresses the conditions under which objects remain at rest, and the laws governing objects in motion.

This chapter covers the basics of mechanics. Chapter 5 covers a specific branch of mechanics—STRENGTH OF MATERIALS.

NEWTON'S LAWS OF MOTION

The study of mechanics is based on Newton's three Laws of Motion:

Newton's First Law states that if a body is at rest, it will continue to be at rest until a force acts upon it. Likewise, if a body is moving at a constant speed and in a straight line, it will continue its movement until a force acts upon it.

Newton's Second Law states that the acceleration (a) of a moving body is directly proportional to the force (F) and inversely proportional to the mass (m) of the body. That is, $a = F/m$ and $F = m \times a$.

Newton's Third Law states that the action of a force upon a body at rest is equal to an opposite force-reaction (static).

FORCE AND MOTION

Force is any action placed on a body that makes it move, changes its motion, or changes its size. Motion is the act, process, or instance of changing place.

A force is a vector quantity characterized by both magnitude and direction. Forces are represented graphically by a line with an arrowhead at one end. The length of the line represents the magnitude of the force, the arrowhead represents its direction, and the end opposite the arrowhead represents the force's application.

The magnitude of a force may be expressed in different units. Commonly used units are the *dyne, kilogram force (kg·f, kgf, or kp), newton (N), pound force (lbf or lb·f),* and *poundal (pdl)*. See Figure 4-1 for the relationship of these units.

Most practical problems related to forces include two or more forces acting simultaneously upon a body. Any two or more forces (F_1, F_2, F_3....F_n) may be replaced by a single force R, called the *resultant*, which has the same effect

FORCE	dyne	kgf	N	lb·f	pdl
1 dyne	1	1.020×10^{-6}	10^{-5}	2.248×10^{-6}	7.233×10^{-5}
1 kilogram	9.807×10^{5}	1	9.807	2.205	70.93
1 newton	10^{5}	0.1020	1	0.2248	7.233
1 pound	4.448×10^{5}	0.4536	4.448	1	32.17
1 poundal	1.383×10^{4}	0.0141	0.1383	0.03108	1

upon the body. Two types of resultants are the *resultant of collinear forces* and the *resultant of concurrent forces.*

Resultant of Collinear Forces

The resultant of collinear forces, that is, forces with vectors that lie along the same straight line, is equal to their algebraic sum. Its direction is always toward the direction of the greater force (Figure 4-2).

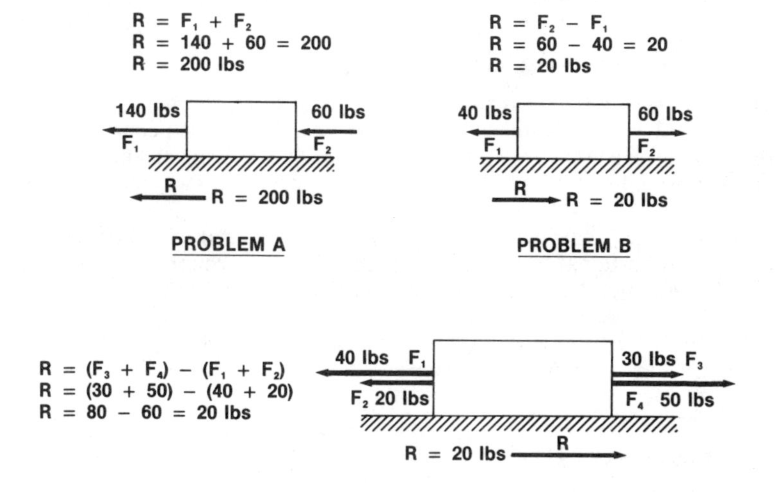

Figure 4-2. Three problems (A, B, and C) showing the resultants of collinear forces.

Resultant of Concurrent Forces

The resultant of concurrent forces, that is, forces with lines of application that pass through a common point, and vectors form an angle. Its direction is always closer to the direction of the greatest force. Concurrent forces are represented by the symbols F_x and F_y. Their resultant can be found graphically through a parallelogram of forces, or it can be found mathematically (Figure 4-3).

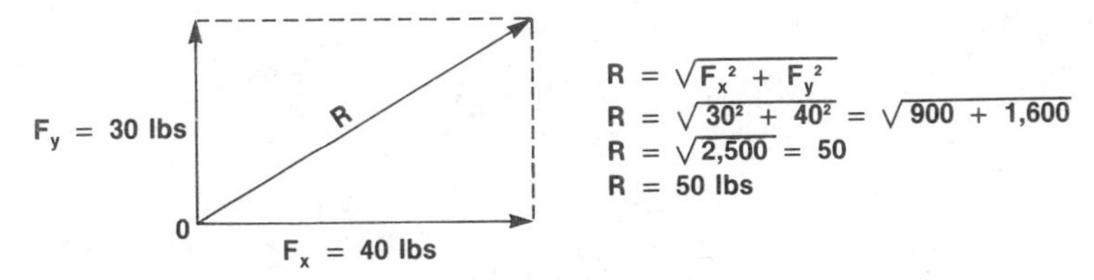

Figure 4-3. Resultant of concurrent forces shown graphically and mathematically.

LINEAR MOTION

Linear motion is the movement of a body in a straight line. The solution of problems related to linear motion requires knowledge of *speed, velocity,* and *acceleration.*

Speed

Speed is the distance per unit of time, or distance traveled in some time interval. Speed is a *scalar quantity*; that is, it is specified by its magnitude only.

Velocity

Velocity is a rate of change in distance per unit of time in a given direction. Velocity is a *vector quantity*; that is, it is specified by both its magnitude and its direction.

Acceleration

Acceleration is a rate of change of velocity per unit of time, or a rate of change of distance per unit of time squared. Acceleration is a vector quantity resulting from a change in either direction or speed.

Units of Linear Motion

Commonly used units of linear motion are *feet-per-second (ft/s), meters-per-second (m/s), kilometers-per-hour (km/h),* and *miles-per-hour (mi/h or MPH).* See Figure 4-4 for the relationship of these units.

SPEED	ft/s	m/s	km/h	mi/h
1 foot per second	1	0.3048	1.097	0.6818
1 meter per second	3.281	1	3.6	2.237
1 kilometer per hour	0.9113	0.2778	1	0.6214
1 mile per hour	1.467	0.4470	1.609	1

Figure 4-4. Relationship between common units of linear motion.

Many problems related to linear motion can be solved by using one of the following formulas:

Linear Motion Formulas

$$a = \frac{V - V_o}{t}, \quad V_m = \frac{V_o + V}{2}, \quad D = V_m \times t$$

a = acceleration in ft/s^2 or m/s^2

V = velocity at the end of time (t) in feet per second (ft/s) or meters per second (m/s)

V_o = initial velocity (zero)

V_m = average velocity

D = distance covered at the end of time (t) in feet or meters (m)

t = time in seconds (s)

See Examples 4-1 and 4-2 for applications of these formulas.

Example 4-1: A train moves with a uniform acceleration from zero to 100 miles per hour in 20 seconds. Find the velocity (V), the acceleration (a), the average velocity (V_m), and the distance (D) covered within 20 seconds.

Given: $V_o = 0$, $V = 100$ MPH, $t = 20$ sec

NOTE: MPH must be converted to ft/s and the hour must be converted to seconds.

Solution: Conversion: 1 ft/s = 0.3048 m/s

$$V = \frac{100 \times 5{,}280}{1 \times 3{,}600} = 147 \text{ ft/s} \qquad 147 \times 0.3048 = 44.8 \text{ m/s}$$

$$a = \frac{V - V_o}{t} = \frac{147 - 0}{20} = 7.35 \text{ ft/s}^2 \qquad 7.35 \times 0.3048 = 2.24 \text{ m/s}^2$$

$$V_m = \frac{V_o + V}{2} = \frac{0 + 147}{2} = 73.5 \text{ ft/s} \qquad 73.5 \times 0.3048 = 22.4 \text{ m/s}$$

$$D = V_m \times t = 73.5 \times 20 = 1{,}470' \qquad 1{,}470 \times 0.3048 = 448 \text{ m}$$

Example 4-2: A train moves with a uniform acceleration from zero to 180 kilometers per hour in 25 seconds. Find the velocity (V), the acceleration (a), the average velocity (V_m), and the distance (D) covered within 25 seconds.

Given: $V_o = 0$, $V = 180$ km/h, $t = 10$ sec

NOTE: km/h must be converted to m/s and the hour must be converted into seconds.

Solution: Conversion: 1 m/s = 3.281 ft/s

$$V = \frac{180 \times 1{,}000}{1 \times 3{,}600} = 50 \text{ m/s} \qquad 50 \times 3.281 = 164 \text{ ft/s}$$

$$a = \frac{V - V_o}{t} = \frac{50 - 0}{25} = 2 \text{ m/s}^2 \qquad 2 \times 3.281 = 6.562 \text{ ft/s}^2$$

$$V_m = \frac{V_o + V}{2} = \frac{0 + 50}{2} = 25 \text{ m/s} \qquad 25 \times 3.281 = 82 \text{ ft/s}$$

$$D = V_m \times t = 25 \times 25 = 625 \text{ m} \qquad 625 \times 3.281 = 2{,}051'$$

ANGULAR MOTION

Angular motion is the movement of a rotating body. The velocity of a rotating body is the distance that a point at its circumference travels per unit of time. This distance is equal to the length of the circumference times the number of rotations (revolutions) per unit of time.

Units of Angular Motion

The units of angular motion are the *degree, minute, second, radian,* and *revolution.* See Figure 4-5 for the relationship of these units.

ANGLE	DEGREES (°)	MINUTES (″)	SECONDS (′)	RADIANS (rad)	REVOLUTION (U)
1 degree	1	60	3,600	0.01745	0.002778
1 minute	0.016667	1	60	2.909×10^{-4}	4.630×10^{-5}
1 second	2.778×10^{-4}	0.01667	1	4.848×10^{-6}	7.716×10^{-7}
1 radian	57.30	3438	206,300	1	0.1592
1 revolution	360	21,600	1,296,000	6.283	1

Figure 4-5. Relationship between common units of angular motion.

Problems related to velocity can be solved by using the following two formulas:

Angular Motion Formulas

$$V = \frac{\pi \times D \times N}{720} = \text{ft/s} \quad \text{or} \quad V = \frac{\pi \times D \times N}{60{,}000} = \text{m/s}$$

$\pi = 3.14$
D = diameter in inches
N = revolutions per minute (RPM)
12 - to convert inches to feet
60 - to convert minutes (min) to seconds (s)
$720 = 12 \times 60$

$\pi = 3.14$
D = diameter in millimeters (mm)
N = revolutions per minute (RPM)
1,000 - to convert millimeters (mm) to meters (m)
60 - to convert minutes (min) to seconds (s)
$60{,}000 = 1{,}000 \times 60$

See Examples 4-3 and 4-4 for applications of these formulas.

Example 4-3: A pulley 8″ in diameter is attached to a rotating shaft. Find its velocity in feet per second (ft/s) when the shaft is turning at a rate of 600 RPM.
Given: D = 8″, N = 600 RPM

Solution: $V = \frac{\pi \times D \times N}{720} = \frac{3.14 \times 8 \times 600}{720} = \frac{15{,}072}{720} = 20.9 \text{ ft/s}$

Example 4-4: A pulley 200 mm in diameter is attached to a rotating shaft. Find its velocity in meters per second (m/s) when the shaft is turning at a rate of 800 RPM.

Given: D = 200 mm, N = 800 RPM

$$\text{Solution: } V = \frac{\pi \times D \times N}{60{,}000} = \frac{3.14 \times 200 \times 800}{60{,}000} = \frac{3.14 \times 2 \times 8}{6}$$

$$= \frac{50.24}{6} = 8.4 \text{ m/s}$$

FRICTION

Friction is the resistance to the motion of one body against another body. That is, it is a force that acts upon the two bodies at the point of their contact. The two types of friction are *static friction* and *kinetic friction.*

The maximum force of static or kinetic friction is a variable quantity. It depends on the normal force (perpendicular weight) between the contacting surfaces and a constant called the *coefficient of friction*. The value of this constant is derived from scientific experiments with various combinations of materials. The value of the *coefficient of static friction* (μ_s) is greater than the value of the *coefficient of kinetic friction* (μ_k). This means that once a solid, which rests on another solid, starts sliding, it requires less force to continue its movement. The formulas for calculating friction are:

Friction Formulas

$$f_s = W \times \mu_s \quad \text{or} \quad f_k = W \times \mu_k$$

f_s = maximum force of static friction
μ_s = coefficient of static friction
f_k = maximum force of kinetic friction
μ_k = coefficient of kinetic friction
W = normal force between contacting surfaces (perpendicular weight)

See Example 4-5 for application of these formulas.

Example 4-5: A 20 lb steel block rests on a flat steel surface. Find the static friction and kinetic friction.

Given: W = 100 lb, $\mu_s = 0.15$, $\mu_k = 0.09$

Solution: $f_s = W \times \mu_s = 100 \times 0.15 = 15$ lbs
$f_k = W \times \mu_k = 100 \times 0.09 = 9$ lbs

The coefficient of friction depends on the type of friction, the combination of the two materials that come in contact, and whether the surfaces are dry or wet (Figure 4-6).

MATERIALS	DRY (μ_s)	DRY (μ_k)	WET (μ_s)	WET (μ_k)
Steel on steel	0.20	0.12	0.15	0.09
Steel on ice	0.03	0.01	—	—
Hemp rope on wood	0.50	0.40	—	—
Leather on oak	0.50	0.30	0.40	0.25
Wrought iron on cast iron or brass	0.31	0.18	0.16	0.10
Tire rubber on concrete	1.00	0.70	0.70	0.50

Figure 4-6. Coefficient of friction [static (μ_s) and kinetic (μ_k)] of selected materials.

WORK

Work (W) in mechanics is the displacement (movement) of an object by a constant force (F) to a specific distance (d). The amount of work done by a constant force depends on the magnitude of the force and its direction, and the distance moved. If the direction of the force is parallel to the movement, as in lifting an object (Figure 4-7), the work done is: W = F × d. If the direction of the force is at an angle (Figure 4-8), the work done is: W = F × d (cos A).

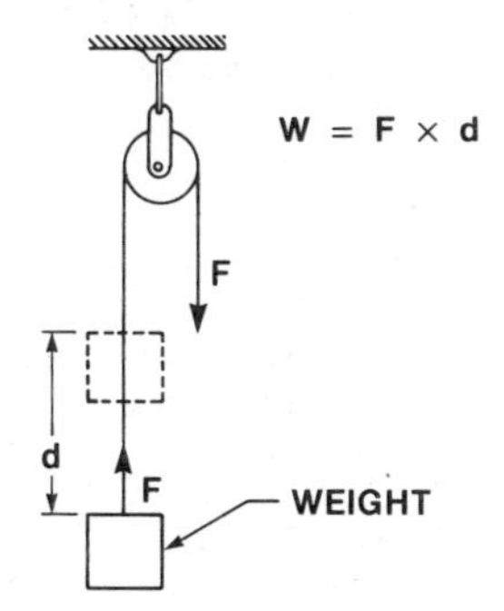

Figure 4-7. Work when lifting an object.

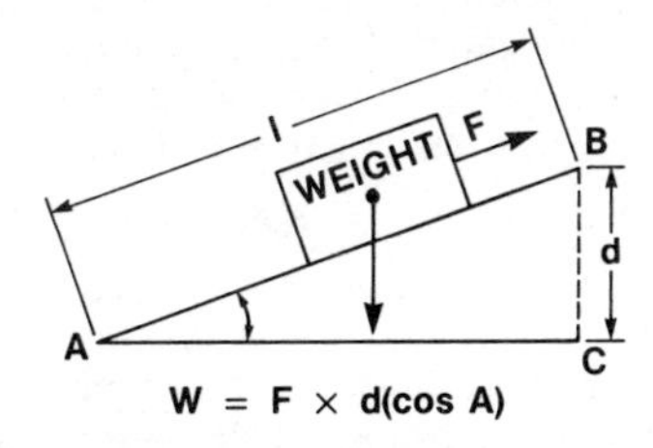

Figure 4-8. Work when sliding an object.

ENERGY

Energy in mechanics is the ability of a falling or moving body (mass and motion) to produce work. Mechanical energy exists in two forms: *potential* and *kinetic*.

Units of Energy

Frequently used units for measuring various forms of energy are the British thermal unit (Btu), the foot-pound (ft.lb.), the Joule (J), the Kilocalorie (Kcal), and the Kilowatt-hour (kWh). See Figure 4-9 for the relationship of these units.

ENERGY	Btu	ft lb	J	kcal	kWh
1 British thermal unit	1	777.9	1,055	0.2520	2.930×10^{-4}
1 foot-pound	1.285×10^{-3}	1	1.356	3.240×10^{-4}	3.766×10^{-7}
1 Joule	9.481×10^{-4}	0.7376	1	2.390×10^{-4}	2.778×10^{-7}
1 kilocalorie	3.968	3,086	4.184	1	1.163×10^{-3}
1 kilowatt-hour	3,413	2.655×10^{6}	3.6×10^{6}	860.2	1

Figure 4-9. Relationship between common units of energy.

The formulas for computing potential energy and kinetic energy are:

Potential and Kinetic Energy Formulas

$$PE = W \times h$$

PE = potential energy in joules (J) or foot pounds (ft lbs)
W = weight in newtons (N) or pounds of force (lb·f)
h = height in meters (m) or feet

$$KE = \frac{m \times V^2}{2}$$

KE = kinetic energy in joules (J) or foot pounds (ft lbs)
m = mass in kilograms (kg) or (slugs)
V = velocity in meters per second (m/s) or feet per second (ft/s)

In addition to mechanical energy, there are several other forms of energy: heat, electrical, magnetic, chemical, and atomic, or nuclear. One form of energy may be changed into another form. For example, electrical energy may be changed into mechanical energy as an electrical motor does.

POWER

Power in mechanics is the rate of work produced per unit of time. Power is expressed as the quotient of work done, divided by the time in which it takes to do the work (Power = work/time). For example, if Motor A can raise an elevator to the third floor of a building in 20 seconds, and Motor B can do the same work in 10 seconds, this means that the power of Motor B is two times greater than the power of motor A.

Units of Power

Commonly used units for measuring power are the *watt (W), foot-pound per second (ft lb/sec), horsepower (HP),* and the *Kilowatt (kW).* See Figure 4-10 for the relationship of these units.

POWER	W	ft lb/s	HP	kW
1 watt	1	0.7376	1.341×10^{-3}	0.001
1 foot-pound/sec	1.356	1	1.818×10^{-3}	1.356×10^{-3}
1 horsepower	745.7	550	1	0.7457
1 kilowatt	1,000	736.6	1.341	1

Figure 4-10. Relationship between common units of power.

EFFICIENCY

Efficiency in mechanics is the ratio of work produced by a machine compared to the energy consumed to produce the work. The work produced is always smaller than the work consumed. The quotient of this ratio determines machine efficiency, and it is expressed as a percentage as shown by this formula:

$$\text{efficiency (e)} = \frac{\text{Power out}}{\text{Power in}} \times 100\%$$

The efficiency of various power-production machines ranges from 15% to 96%:

Electrical motors	80%–96%	Hydraulic turbines	85%–90%
Diesel engines	32%–38%	Steam turbines	80%–85%
Gasoline engines	24%–27%	Steam engines	15%–25%

SIMPLE MACHINES

Simple machines are devices that transfer a force from one point of application to another point of application. Simple machines are used to gain a *mechanical advantage* or to do work with as little effort as possible. Common simple machines are the *lever, pulley, drum,* and *inclined plane.*

Lever

The lever consists of a rigid body that pivots on a fixed body (fulcrum). Every lever has two *moments.* A moment is the product of the applied load times the distance from the fulcrum.

Classes of Levers. There are three classes of levers—first, second, and third. Class is determined by the position of the fulcrum in relation to the applied load (W) and to the force or effort (F). See Figures 4-11, 4-12, and 4-13 for the force or effort required to balance a load of 120 pounds in each class of levers.

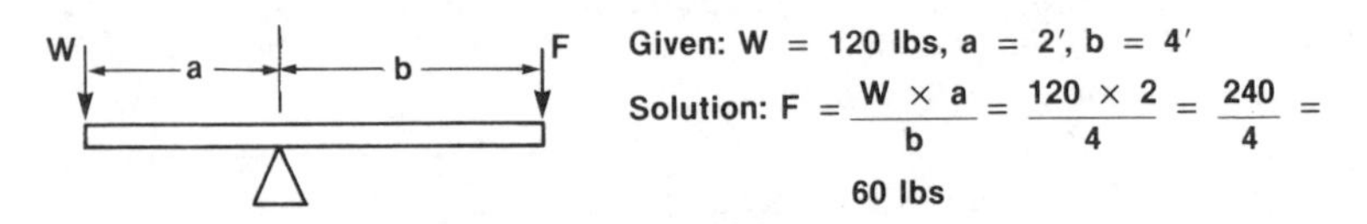

Figure 4-11. First class lever and an example of determining force or effort (F).

W
a
F
b

Given: W = 120 lbs, a = 2′, b = 6′

Solution: $F = \frac{W \times a}{b} = \frac{120 \times 2}{6} = \frac{240}{6} = 40$ lbs

Figure 4-12. Second class lever and an example of determining force or effort (F).

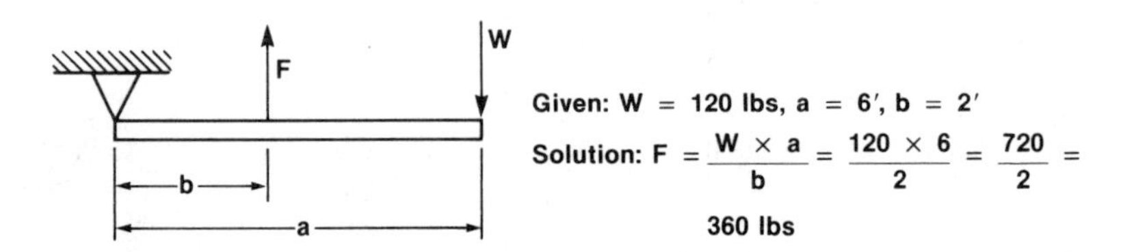

Figure 4-13. Third class lever and an example of determining force or effort (F).

Mechanical Advantage of Levers. The *mechanical advantage* of a machine is the ratio of the resistance to the effort. For instance, if the required effort for lifting a 100-lb weight is 50 lbs, the mechanical advantage is two.

The mechanical advantage of a first class lever may be smaller, greater, or equal to one. The mechanical advantage of a second class lever is greater than one; and the mechanical advantage of a third class lever is smaller than one.

NOTE: Although the load in each class of lever shown in Figures 4-11, 4-12, and 4-13 remains the same, the force (F) required to achieve equilibrium is different.

Pulleys

The pulley is a sheave, or small wheel, with a grooved rim in which a band, belt, or rope passes over. It is used for lifting heavy objects. The mechanical advantage of a pulley is determined by the distance the force or effort has to move in order to lift the load. Four types of pulleys are the *single pulley, double pulley, multi-pulley,* and *differential pulley*. Each has its own mechanical advantage.

The mechanical advantage of a single fixed pulley (Figure 4-14) is one. The mechanical advantage of a double pulley (Figure 4-15) is two. The mechanical advantage of the multi-pulley (Figure 4-16) is equal to the number of pulleys.

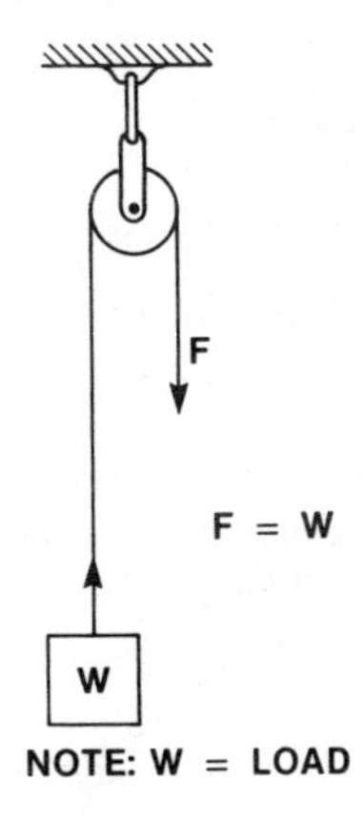

Figure 4-14. Single fixed pulley and force (F) formula.

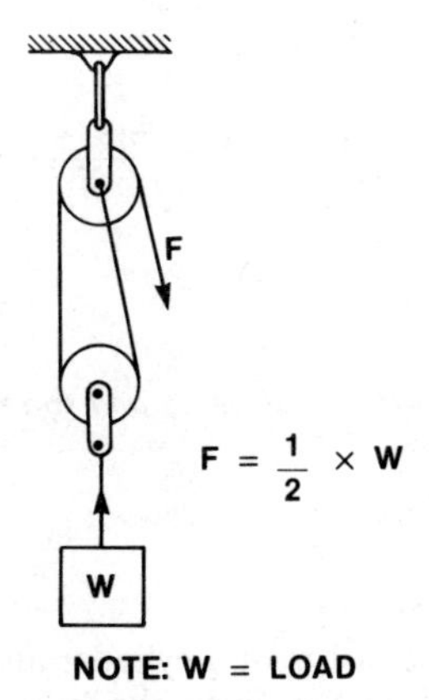

Figure 4-15. Double pulley and force (F) formula.

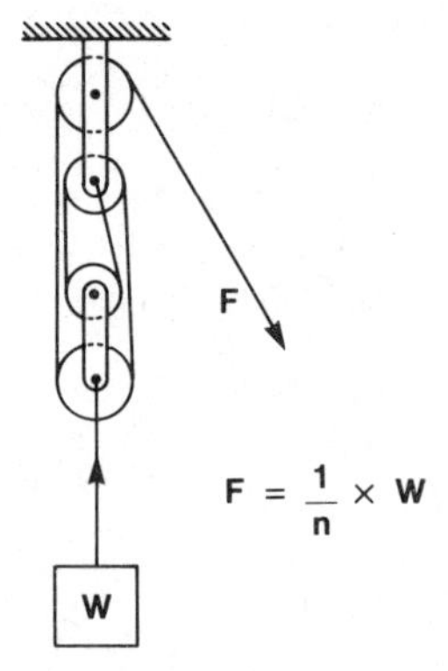

Figure 4-16. Multi-pulley and force (F) formula.

The differential pulley (Figure 4-17) is designed for heavy loads. It uses a chain instead of a rope. The chain engages sprockets to prevent slipping. The mechanical advantage of differential pulleys depends upon the ratio:

$$\frac{R - r}{2R}$$

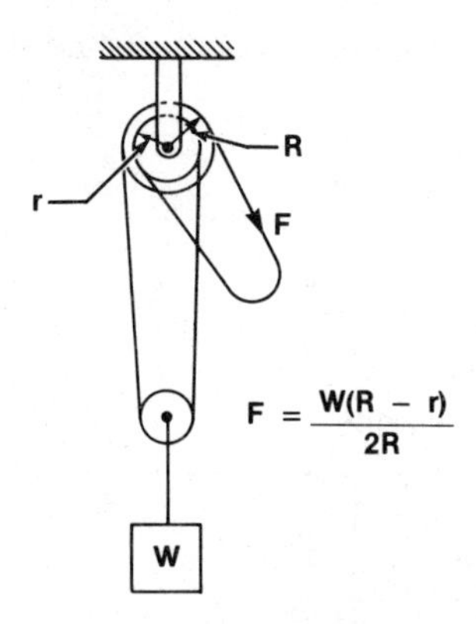

Figure 4-17. Differential pulley and force (F) formula.

Drum (Wheel) and Axle

The drum (wheel) and axle consists of a cylinder and a crank attached to the axle (shaft) of the cylinder. Like the pulley, the drum is used for lifting weights. The mechanical advantage of a drum is equal to the ratio $\frac{r}{R}$. See Figure 4-18.

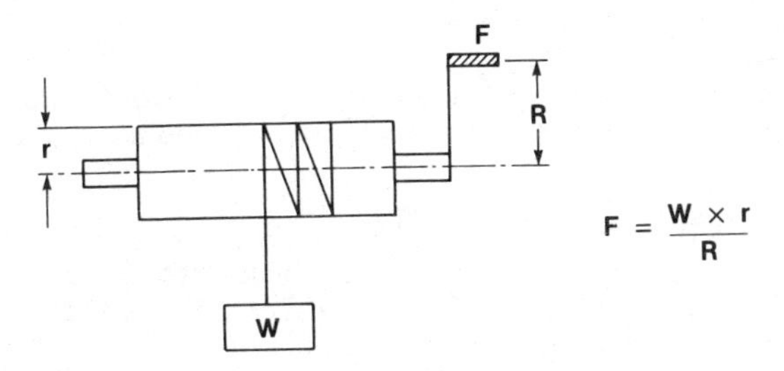

Figure 4-18. Drum (wheel) and axle and force (F) formula.

Inclined Plane

The inclined plane consists of a flat surface, usually a board, one end of which is placed at a higher level than the other. Inclined planes are used to raise heavy objects to a platform or to lower them. The force or effort (F) needed to raise a load (W) to a height (h) is computed by using the formula in Figure 4-19. The force or effort needed to lower a load is computed by using the formula in Figure 4-20. The letters (l) and (μ) in both formulas represent the length of the inclined plane and the coefficient of friction, respectively.

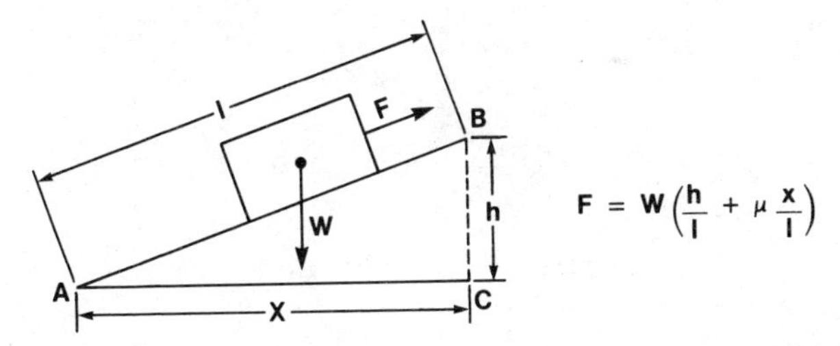

$$F = W\left(\frac{h}{l} + \mu\frac{x}{l}\right)$$

Figure 4-19. Inclined plane used for elevation and its force (F) formula.

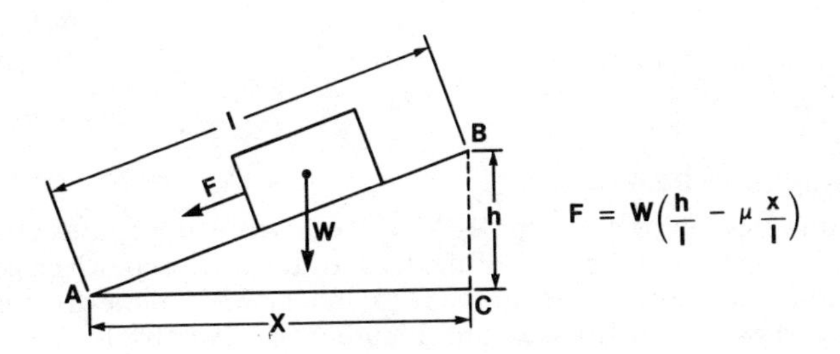

$$F = W\left(\frac{h}{l} - \mu\frac{x}{l}\right)$$

Figure 4-20. Inclined plane used for lowering an object and its force (F) formula.

TRANSMISSION

Transmission in Physical Science refers to the transfer of rotational motion from one shaft to another shaft by means of belts and pulleys, or by gears.

Transmission by Belts and Pulleys

In any transmission using belts and pulleys, the number of revolution-per-minute (RPM) of the driven shaft depends on the RPM of the drive shaft and the ratio of the diameters of the drive and driven pulleys. The relationships between the diameters of the pulleys and the RPM of the drive and driven shaft in a single transmission system and multiple transmission system are shown in Figure 4-21 and Figure 4-22, respectively.

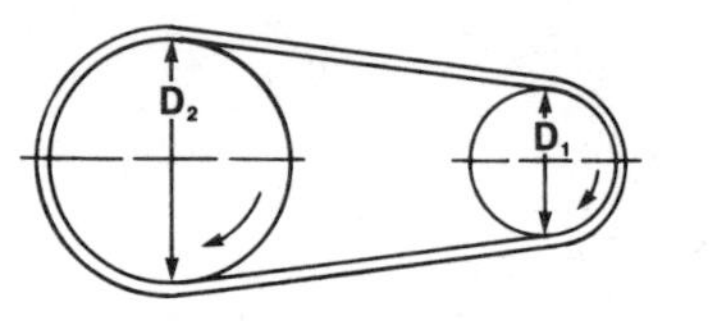

$$N_1 \times D_1 = N_2 \times D_2 \text{ and } N_2 = \frac{N_1 \times D_1}{D_2}$$

N_1 = RPM OF DRIVE SHAFT
D_1 = DIAMETER OF DRIVE PULLEY
N_2 = RPM OF DRIVEN SHAFT
D_2 = DIAMETER OF DRIVEN PULLEY

Figure 4-21. Single transmission system by belt and pulleys and related formulas.

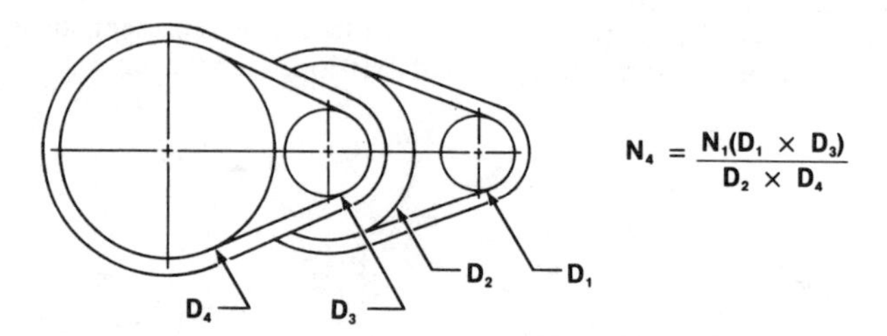

Figure 4-22. Multiple transmission system by belts and pulleys and related formula.

Transmission by Gears

In any transmission by gears, the RPM of the driven shaft depends on the RPM of the drive shaft and the ratio of the teeth of the drive and driven gears. The relationships between the teeth of the gears and the RPM of the drive and driven gears are shown in Figure 4-23 and Figure 4-24, respectively.

The RPM of a driven shaft in any transmission system can be found by using the proper formula. See Examples 4-6 and 4-7.

$N_1 \times G_1 = N_2 \times G_2$ and $N_2 = \dfrac{N_1 \times G_1}{G_2}$

N_1 = RPM OF DRIVE SHAFT
G_1 = NUMBER OF TEETH OF DRIVE GEAR
N_2 = RPM OF DRIVEN SHAFT
G_2 = NUMBER OF TEETH OF DRIVEN GEAR

G_1

G_2

Figure 4-23. Single transmission system by gears and related formulas.

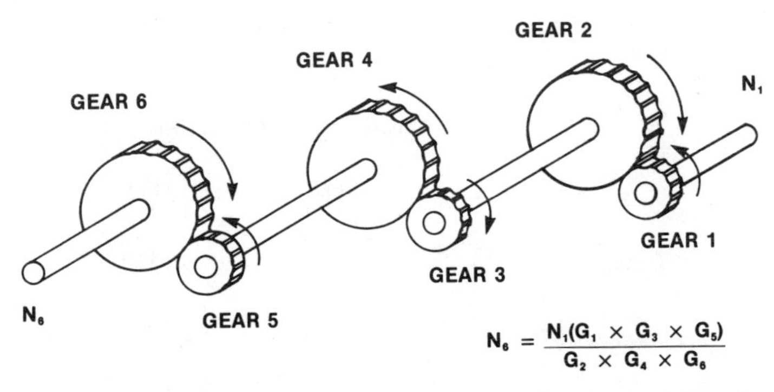

Figure 4-24. Multiple transmission system by gears and related formula.

Example 4-6: In a single transmission system using belts and pulleys, the speed of the drive shaft is 1,200 RPM and the diameter of the pulley attached to it is 4″. Find the RPM of the driven shaft if the diameter of its pulley is 10″.

Given: $N_1 = 1{,}200$ RPM, $D_1 = 4''$, $D_2 = 10''$

Solution: $N_2 = \dfrac{N_1 \times D_1}{D_2} = \dfrac{1{,}200 \times 4}{10} = \dfrac{4{,}800}{10} = 480$ RPM

Example 4-7: In a multiple transmission system (gear train with six gears), the speed of the drive shaft is 6,000 RPM and the gear attached to it has 24 teeth. Find the RPM of the driven shaft if the number of teeth of the gear attached to this shaft is 56.

Given: $N_1 = 6{,}000$ RPM, $G_1 = 24$, $G_2 = 48$, $G_3 = 20$, $G_4 = 60$
$G_5 = 28$, $G_6 = 56$

Solution: $N_6 = \dfrac{N_1(G_1 \times G_3 \times G_5)}{G_2 \times G_4 \times G_6} = \dfrac{6{,}000(24 \times 20 \times 28)}{48 \times 60 \times 56}$

$= \dfrac{6{,}000 \times 13{,}440}{161{,}280} = \dfrac{80{,}640{,}000}{161{,}280} = 500$ RPM

TEMPERATURE AND HEAT (THERMAL ENERGY)

The terms temperature and heat are frequently misused as interchangeable terms. Temperature and heat are related, but they are two distinctly different quantities. Temperature is a measure of the coldness or the hotness of a substance. Heat (thermal energy) is a measure of the internal energy that flows from one substance to another substance.

Units of Temperature

Temperature is measured in *degrees Kelvin (°K), degrees Celsius (°C),* or *degrees Fahrenheit (°F).* Temperature expressed in degrees Kelvin is called *absolute temperature* because its starting point is the *absolute zero*, which is the equivalent of −273.16 °C or −459.69 °F. The measurement of temperature in °K, °C, or °F, is obtained by using one of the three temperature scales shown in Figure 4-25.

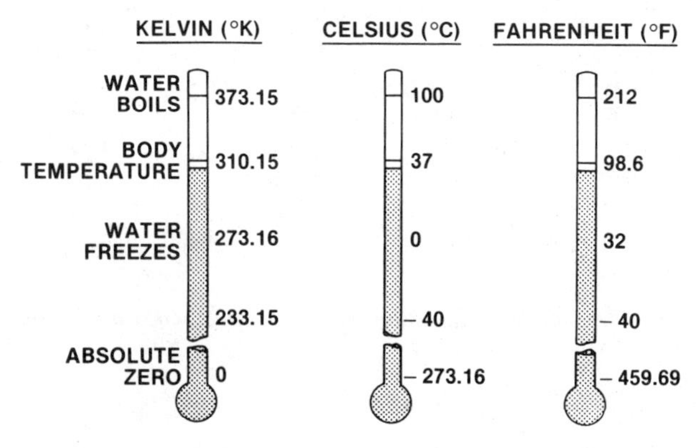

Figure 4-25. Temperature scales and their relationships.

The conversion of temperature units from one scale to another can be done by applying the following formulas:

Temperature Conversion Formulas

Temperature in degrees Kelvin: $t\,°K = t\,°C + 273.16$

Temperature in degrees Celsius: $t\,°C = \frac{5(t\,°F - 32)}{9}$

Temperature in degrees Fahrenheit: $t\,°F = \frac{9 \times t\,°C}{5} + 32$

Examples 4-8 and 4-9 show how to convert temperature units.

The equivalents of temperatures expressed in degrees Celsius and degrees Fahrenheit can also be found in tables (Figure 4-26).

Example 4-8: Convert 72 degrees Fahrenheit (72 °F) to its equivalent temperature in degrees Celsius.
Given: t = 72 °F

Solution: $t\,°C = \frac{5(t\,°F - 32)}{9}$

$$t\,°C = \frac{5(72 - 32)}{9} = \frac{200}{9} = 22.2\,°C$$

Example 4-9: Convert 37 degrees Celsius (37 °C) to its equivalent temperature in degrees Fahrenheit.
Given: t = 37 °C

Solution: $t\,°F = \frac{9 \times t\,°C}{5} + 32$

$$t\,°F = \frac{9 \times 37}{5} + 32 = \frac{333}{5} + 32 = 66.6 + 32 = 98.6\,°F$$

°F	°C	°F	°C	°F	°C	°F	°C	°F	°C
− 160	− 107	340	171	840	449	1340	727	1840	1004
− 140	− 96	360	182	860	460	1360	738	1860	1016
− 120	− 84	380	193	880	471	1380	749	1880	1027
− 100	− 73	400	204	900	482	1400	760	1900	1038
− 80	− 62	420	216	920	493	1420	771	1920	1049
− 60	− 51	440	227	940	504	1440	782	1940	1060
− 40	− 40	460	238	960	516	1460	793	1960	1071
− 20	− 29	480	249	980	527	1480	804	1980	1082
0	− 18	500	260	1000	538	1500	816	2000	1093
20	− 7	520	271	1020	549	1520	827	2020	1104
40	4	540	282	1040	560	1540	838	2040	1116
60	16	560	293	1060	571	1560	849	2060	1127
80	27	580	304	1080	582	1580	860	2080	1138
100	38	600	316	1100	593	1600	871	2100	1149
120	49	620	327	1120	604	1620	882	2120	1160
140	60	640	338	1140	616	1640	893	2140	1171
160	71	660	349	1160	627	1660	904	2160	1182
180	82	680	360	1180	638	1680	916	2180	1193
200	93	700	371	1200	649	1700	927	2200	1204
220	104	720	382	1220	660	1720	938	2220	1216
240	116	740	393	1240	671	1740	949	2240	1227
260	127	760	404	1260	682	1760	960	2260	1238
280	138	780	416	1280	693	1780	971	2280	1249
300	149	800	427	1300	704	1800	982	2300	1260
320	160	820	438	1320	716	1820	993	2320	1271

Figure 4-26. Temperature equivalents.

Units of Heat or Thermal Energy

Heat is measured in *British thermal units (Btu)* or in *kilocalories (kcal)*. The British thermal unit is the quantity of heat required to raise the temperature of 1 lb of water 1 °F. The kilocalorie is the quantity of heat required to raise the temperature of one kilogram (1 kg) of water 1 °C. (For the relationship of these units, refer back to Figure 4-9).

Thermal Energy of Common Fuels

The burning of certain substances produces large amounts of heat which is used for commercial purposes. The heat of these substances (Figures 4-27 and 4-28) is measured in British thermal units per pound (Btu/lb) or British thermal units per cubic foot (Btu/ft^3), and kilocalories per kilogram (kcal/kg) or kilocalories per cubic meter (kcal/m^3).

SOLID FUELS	kcal/kg	Btu/lb	LIQUID FUELS	kcal/kg	Btu/lb
Anthracite	7,500–8,500	13,500–15,300	Gasoline	10,400–11,400	18,700–20,500
Briquettes	6,500–8,000	11,700–14,500	Kerosine	10,500–11,200	18,900–20,200
Coke	6,800–7,200	12,300–13,000	Diesel oil	10,000–10,500	18,000–19,000
Lignite	2,500–5,000	4,500–9,000	Heating oil	9,800–11,000	17,600–19,800
Wood	2,800–3,600	5,000–6,500	Alcohol	6,400–7,100	11,500–12,800

Figure 4-27. Thermal energy of selected ore fuels.

Acetylene	13,500 kcal/m^3 or 1,500 Btu/ft^3
Natural gas	9,800 kcal/m^3 or 1,100 Btu/ft^3
Methane	8,500 kcal/m^3 or 950 Btu/ft^3

Figure 4-28. Thermal energy of selected gas fuels.

Specific Heat Capacity

Specific heat capacity (c) of a substance is the quantity of heat required to increase the temperature of a unit mass of the substance by one degree. The specific heat capacity of a substance is a constant, the numerical value of which depends on the substance (Figure 4-29).

METALLIC SOLIDS		NONMETALLIC SOLIDS		LIQUIDS	
Aluminum	0.212	Ice	0.430	Water	1.000
Brass	0.090	Clay	0.220	Ethyl alcohol	0.580
Copper	0.094	Coal	0.300	Gasoline	0.500
Gold	0.031	Concrete	0.160	Mercury	0.033
Iron and steel	0.110	Glass	0.1–0.2	Mineral oil	0.500
Lead	0.031	Limestone	0.220	Methyl alcohol	0.600
Platinum	0.032	Marble	0.210	Olive oil	0.470
Silver	0.056	Paraffin	0.690	Petroleum	0.510
Tin	0.055	Rubber	0.480	Sea water	0.930
Zinc	0.094	Wood	0.3–0.7	Turpentine	0.410

Figure 4-29. Specific heat capacity of selected substances.

The amount of heat required to raise the temperature of a mass (m) of a substance depends on its specific heat capacity (c) and the change of temperature from its original temperature (t_o) to its final temperature (t_i). The formula for computing this is:

Heat, Mass, Temperature Formula

$$Q = m \times c(Td)$$

$Q = \text{Btu}$	$Q = \text{kcal}$
$c = \text{Btu/lb} \times {}^\circ\text{F}$	$c = \text{kcal/kg} \times {}^\circ\text{C}$ or $\text{cal/g} \times {}^\circ\text{C}$
$Td = {}^\circ\text{F}$	$Td = {}^\circ\text{C}$

NOTE: Td is found by subtracting t_i from t_o.

See Example 4-10 for an application of this formula.

Example 4-10: A ten kilogram steel shaft is to be heat treated at 475 °C. Find the amount of heat needed if the original temperature of the shaft is 25 °C.

Given: $m = 10$ kg, $t_i = 475\,^\circ\text{C}$, $t_o = 25\,^\circ\text{C}$, $C = 0.11$ kcal/ °C

Solution: $Q = m \times c(Td) = 10 \times 0.11(450) = 495$ kcal

Heat Rate

Heat rate (Q_r) is the amount of heat (Q) supplied by a heating element per unit time, usually per minute. The time required to raise or lower the temperature of a substance depends on the heat needed and the heat rate supplied by the heating element. The formula for computing this is:

Heat Rate Formula

$$t = \frac{Q}{Q_r}$$

t = time in minutes
Q = heat in Btu or kcal
Q_r = heat rate in Btu/min or kcal/min

See Example 4-11 for an application of this formula.

Example 4-11: A 20-pound aluminum mass is to be melted at 1,220 °F. Find the amount of heat (Q) needed if the original temperature of the aluminum mass is 70 °F, and the time (t) required for the melting if the heating element supplies heat at the rate of 500 Btu/min.

Given: $m = 20$ lbs, $t_i = 1{,}220\,^\circ\text{F}$, $t_o = 70\,^\circ\text{F}$, $c = 0.212$,
$Q_r = 500$ Btu/min

Solution: $Q = m \times c(t_i - t_o) = 20 \times 0.212(1{,}150) = 4{,}876$ Btu

$$t = \frac{Q}{Q_r} = \frac{4{,}876}{500} = 9.75 \text{ min}$$

Fuel Consumption

The quantity of fuel required to produce a certain amount of heat depends on the type of fuel used. This quantity can be computed by using the formula:

Fuel Consumption Formula

$$F = \frac{Q}{E}$$

F = fuel in kg, lbs, m^3, or ft^3
Q = amount of heat in kcal or Btu
E = thermal energy of the fuel in $\frac{kcal}{kg}$, $\frac{Btu}{lbs}$, $\frac{kcal}{m^3}$, or $\frac{Btu}{ft^3}$

See Example 4-12 for an application of this formula.

Example 4-12: The heating element in Example 4-11 uses a gas fuel with a thermal energy of 1,100 Btu/ft^3. Find the amount of gas consumed to produce the 4,876 Btu needed to melt the 20 lbs of aluminum.

Given: Q = 4,876 Btu, E = 1,100 Btu/ft^3

Solution: $F = \frac{Q}{E} = \frac{4{,}876}{1{,}100} = 4.43\ ft^3$

DENSITY AND SPECIFIC GRAVITY

Density and specific gravity are properties that characterize all materials. Each material has its own density and specific gravity (Figure 4-30).

Density

Density is the ratio of a portion of mass (weight) to the volume of the same portion. It is usually expressed in lbs/ft^3, lbs/in^3, or kg/m^3, kg/dm^3, and g/cm^3.

Specific Gravity

Specific gravity is a constant that indicates the relationship between the weight of a certain volume of a solid or liquid material, to the same volume of water at 60 °F or 16 °C; or it indicates the weight of a certain volume of gas to the same volume of air at 32 °F or 0 °C. The density and the specific gravity of any substance can be found by using the formula:

Specific Gravity Formulas

$$\text{Specific Gravity (SP. GR.) for Solids-Liquids} = \frac{\text{Density of a unit volume of solid or liquid}}{\text{Density of the same unit volume of water}}$$

$$\text{Specific Gravity (SP. GR.) for Gases} = \frac{\text{Density of a unit volume of gas}}{\text{Density of the same unit volume of air}}$$

See Examples 4-13 and 4-14 for specific applications.

SOLIDS	SPECIFIC GRAVITY	DENSITY			
		lbs/in³	lbs/ft³	g/cm³	kg/dm³
Aluminum	2.700	0.0975	163.5	2.7	2.7
Antimony	6.618	0.2390	413.0	6.613	6.613
Copper	8.890	0.3210	554.7	8.890	8.890
Gold	19.300	0.6969	1204.3	19.300	19.300
Iron	7.860	0.2841	491.0	7.860	7.860
Lead	11.842	0.4895	707.7	11.342	11.342
Nickel	8.800	0.3177	549.1	8.800	8.800
Silver	10.530	0.3807	657.1	10.530	10.530
Tin	7.290	0.2632	454.9	7.290	7.290
Zinc	7.160	0.2585	446.8	7.160	7.160
LIQUIDS					
Alcohol	0.790	0.00285	49.28	0.790	0.790
Ammonia	0.890	0.03212	55.59	0.890	0.890
Gasoline	0.700	0.0253	43.67	0.700	0.700
Kerosene	0.800	0.0289	49.91	0.800	0.800
Naphtha	0.760	0.0274	47.41	0.760	0.760
Oil (mineral)	0.920	0.0332	57.39	0.920	0.920
Oil (olive)	0.920	0.0332	57.99	0.920	0.920
Oil (petroleum)	0.820	0.0296	51.15	0.820	0.820
Water	1.000	0.0361	62.38	1.000	1.000
Water (sea)	1.030	0.0372	64.25	1.030	1.030
GASES					
Acetylene	0.907	—	14.75	—	—
Air	1.000	—	13.26	—	—
Argon	1.379	—	9.5	—	—
Helium	0.1381	—	95.90	—	—
Hydrogen	0.0695	—	191.50	—	—
Natural gas	0.667	—	19.75	—	—
Oxygen	1.105	—	12.00	—	—

Figure 4-30. Specific gravity and density of selected materials.

Example 4-13: The weight of one cubic foot of aluminum is 168.5 lbs. Find its specific gravity (SP.GR.).

Given: Density of aluminum - 168.50 lbs/ft³
Density of water - 62.41 lbs/ft³

Solution: $\text{SP.GR. of Aluminum} = \dfrac{\text{weight of aluminum}}{\text{weight of water}} = \dfrac{168.50}{62.41} = 2.69$

Example 4-14: The specific gravity of oxygen is 1.106. Find its weight per cubic foot if the density of air is 0.0807 lbs/ft³.

Given: SP.GR. of oxygen = 1.106, Density of air = 0.0807 lbs/ft³

Solution: Density of oxygen = SP.GR. of oxygen × Density of air

Density of oxygen = 1.106 × 0.0807 = 0.089254 lbs/ft³

FLUID AND PRESSURE

The term fluid is often used when referring to liquid and gaseous materials. Although liquids are incompressible and gases are compressible, all fluid materials assume the shape of their container and exert uniform pressure to all points of their container. For deep containers, the liquid pressure is proportional to the depth.

Pressure

Pressure is the force per unit area. That is, pressure depends not only on the magnitude of the force that causes it, but it also depends on the size of the cross-sectional area upon which the force acts. The formula for finding pressure is:

Pressure Formula

$$P = \frac{F}{A},\ F = P \times A,\ A = \frac{F}{P}$$

P = pressure in lbs/in^2 or Pascals (Pa)
F = force in lb·f or in Newtons (N)
A = area in in^2 or in mm^2

According to *Pascal's Principle*, pressure applied to an enclosed liquid is transmitted uniformly in all directions to every portion of the liquid and to the walls of the container (Figure 4-31).

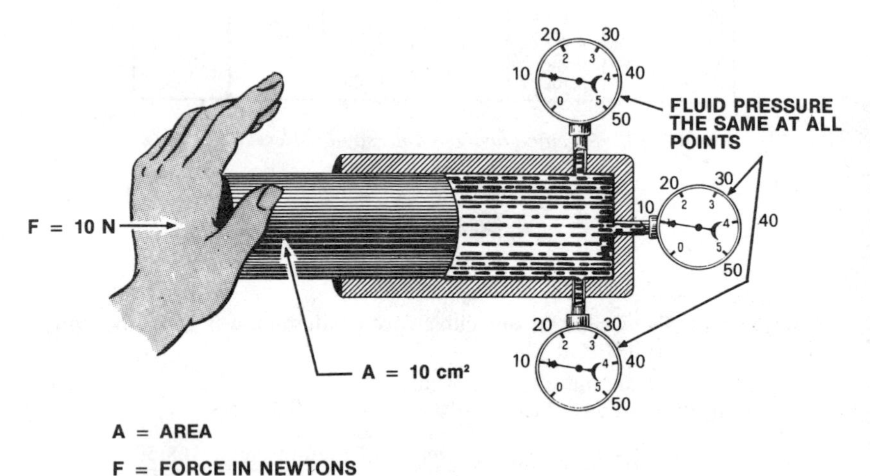

Figure 4-31. An example of Pascal's Principle (law). (Sun Company)

Hydraulic Press

Pascal's Principle concerning the incompressibility of liquids is used to either increase the pressure or to increase the force, as shown by the hydraulic press in Figure 4-32.

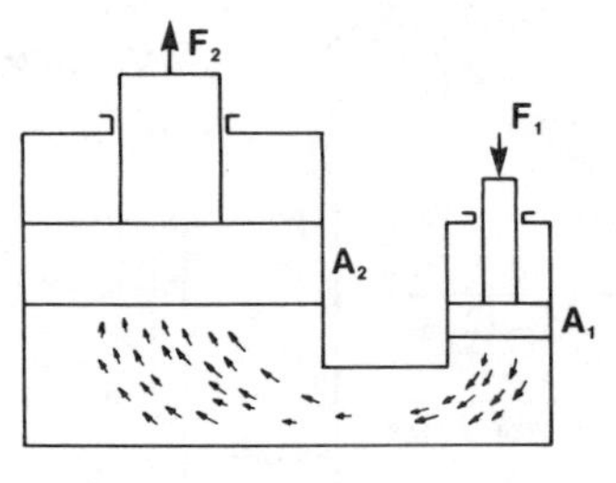

Figure 4-32. This example of a hydraulic press demonstrates Pascal's Principle concerning the incompressibility of liquids.

In any hydraulic press, if a force F_1 is applied to the area A_1 of the small piston, it would create pressure P in the fluid inside the cylinder. This pressure is then transferred into the large cylinder, pushing its piston with force F_2. See Example 4-15.

Example 4-15: In a hydraulic press, the input force (F_1) is 10,000 lbs. Find the pressure (P) in pounds per square inch (lbs/in²) and pascals (Pa). Then find the force (F_2) in pounds (lbs) and newtons (N).

Given: $F_1 = 10{,}000$ lb·f $A_1 = 20$ in², $A_2 = 80$ in²

Conversion factors - 1 lb·f = 4.45 N
1 lb/in² = 6.895 kPa

Solution: $P = \frac{F_1}{A_1} = \frac{10{,}000}{20} = 500$ lbs/in² or $500 \times 6.895 = 3{,}447.5$ kPa

or 3.4475 MPa

$F_2 = P \times A_2 = 500 \times 80 = 40{,}000$ lb·f or $40{,}000 \times 4.45$
$= 178{,}000$ N $= 178$ kN

Atmospheric Pressure

The air in the atmosphere has a definite weight, and because of this weight, everything on the surface of the earth is subjected to *atmospheric pressure.*

The effect of atmospheric pressure is shown on the manometer in Figure 4-33 and on the barometer in Figure 4-34. In both the manometer and the barometer, the atmospheric pressure supports a column of mercury inside a vacuum glass tube, the height of which is an indicator of the magnitude of this pressure.

Atmospheric pressure varies according to altitude, location, and time. However, at sea level, when the temperature is 0 °C or 32 °F, atmospheric pressure is always the same. This pressure that corresponds to 76 centimeters, or 29.92″, of mercury is taken as a unit. This unit is called normal *atmosphere (atm)* and is used as reference point (0) for the calibration of pressure gauges. See Figure 4-35 for the equivalents of one atmosphere (1 atm).

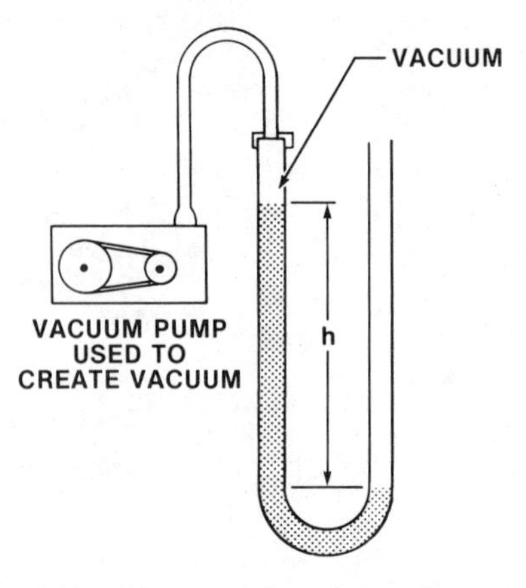

Figure 4-33. Manometer principles of operation.

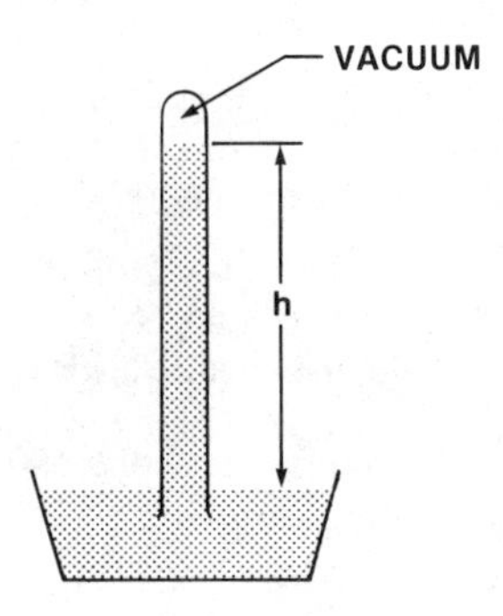

Figure 4-34. Barometer principles of operation.

1 atm IS EQUAL TO:
76 centimeters of mercury (76 cm·Hg)
29.92 inches of mercury (29.92 in·Hg)
406.8 inches of water (406.8 in·H_2O)
39.9 feet of water (39.9 ft·H_2O)
10.33 meters of water (10.33 m·H_2O)
14.70 pounds per square inch (14.70 lb·f/in^2)
2,116 pounds per square foot (2,116 lb·f/ft^2)
101.325 newtons per square meter (101.325 N/m^2)

Figure 4-35. Equivalents of one atmosphere (1 atm).

Common Pressure Units

Common pressure units are: the *atmosphere*, the *inch of water*, the *centimeter of mercury*, the *Pascal*, or *Newton per square meter*, and the *pound per square inch*. See Figure 4-36 for the relationship of these units.

PRESSURE UNITS	atm	in·H_2O	cm·Hg	Pa or N/m^2	lb/in^2 or psi
1 atmosphere atm	1	406.8	78	1.013×10^5	14.70
1 inch of water in H_2O (at 4°C)	0.00246	1	0.1868	249.1	0.03613
1 centimeter of mercury cm·Hg (at 0°C)	0.01316	5.353	1	1,333	0.1934
1 Pascal or newton per square meter Pa or N/m^2	9.869×10^{-6}	0.00411	7.501×10^{-4}	1	1.450×10^{-4}
1 pound per square inch lb/in^2 or psi	0.06805	27.68	5.171	6,895	1

Figure 4-36. Relationship between common pressure units.

Other units of pressure used in various applications are: the *bar,* or *dyne per centimeter square (dyne/cm²)*; the *torr,* or *millimeter of mercury (mm·Hg)*; the *kilogram force per square centimeter (kg·f/cm²);* and the *pound force per square foot (lb·f/ft²).*

The relationship between these units of pressure and the kilopascal (kPa) or 1,000 Pa is shown in Figure 4-37.

1 bar or 1 dyne/cm^2	=	1 kPa (kilopascal)
1 torr or 1 mm·Hg	=	1.3332 kPa
1 kg·f/cm^2	=	9.8066 kPa
1 lb·f/ft^2	=	4.7880 kPa

Figure 4-37. Relationship of pressure units to the kilopascal (kPa).

GAS LAWS

Gases are compressible and they exert uniform pressure at every point of their container. They also assume the shape of their container. Because gases are compressible, a large quantity of any gas may be stored in a relatively small container. The laws that apply to gases are *Boyle's Law, Charles' Law,* and the *Ideal Gas Law.*

Boyle's Law

Boyle's Law states that the volume of a fixed quantity of gas varies inversely with the pressure if the temperature is constant. The formula derived from this law is:

Boyle's Law Formula

$$\frac{V_1}{V_2} = \frac{P_2}{P_1} \text{ or } P_1 \times V_1 = P_2 \times V_2$$

P_1 = initial pressure
V_1 = initial volume in in^3 or mm^3

P_2 = change in pressure
V_2 = change in volume in in^3 or mm^3

See Example 4-16 for an application of this formula.

Example 4-16: Find the change in pressure P_2 of a compressed gas in a cylinder and convert this pressure into metric units (Pa, kPa, and MPa).
Given: initial pressure P_1 = 200 lbs/in^2, initial volume V_1 = 100 in^3
change in volume V_2 = 20 in^3,
Conversion factors - 1 lb/in^2 = 6,895 Pa
kPa = Pa^3, MPa = Pa^6
Solution: $P_1 \times V_1 = P_2 \times V_2$ and

$$P_2 = \frac{P_1 \times V_1}{V_2} = \frac{200 \times 100}{20} = 1{,}000 \text{ lb/in}^2$$

$1{,}000 \times 6{,}895 = 6{,}895{,}000$ Pa
6,895,000 Pa = 6,895 kPa or 6.895 MPa

Charles' Law

Charles' Law states that the volume of a definite quantity of confined gas varies directly with the absolute temperature if the pressure is constant. The formula derived from Charles' Law is:

Charles' Law Formula

$$\frac{V_1}{T_1} = \frac{V_2}{T_2} \text{ or } V_1 \times T_2 = V_2 \times T_1$$

V_1 = initial volume
T_1 = initial absolute temperature in °K
V_2 = change in volume
T_2 = final absolute temperature in °K

NOTE: The temperature in Charles' Law formula must be expressed in degrees Kelvin or absolute temperature. Zero degrees Kelvin (0 °K) corresponds to −273.16 °C or −459.67 °F. (See Chapter 2, *Temperature Scales and Their Relationship*, for additional information.)

See Example 4-17 for a typical application.

Example 4-17: Find the change in volume V_2 of a gas due to change in temperature.

Given: initial volume $V_1 = 150$ lbs/in³, initial temperature $t_1 = 100\,°F$, final temperature $t_2 = 200\,°F$

NOTE: To determine the absolute temperatures T_1 and T_2:

$$T_1 = t_1 + 459.67 = 100 + 459.67 = 559.67\,°K$$
$$T_2 = t_2 + 459.67 = 200 + 459.67 = 659.67\,°K$$

Solution: $\dfrac{V_1}{T_1} = \dfrac{V_2}{T_2}$ and $V_2 = \dfrac{V_1 \times T_2}{T_1}$

$$V_2 = \frac{150 \times 659.67}{559.67} = \frac{98{,}950.5}{559.67} = 176.80 \text{ in}^3$$

Ideal Gas Law

The Ideal Gas Law combines Boyle's Law and Charles' Law into a single law. The formula derived from the Ideal Gas Law is:

Ideal Gas Law Formula

$$\frac{P_1 \times V_1}{T_1} = \frac{P_2 \times V_2}{T_2}$$

P_1 = initial pressure
V_1 = initial volume
T_1 = initial absolute temperature
P_2 = change in pressure
V_2 = change in volume
T_2 = change in absolute temperature

See Example 4-18 for a typical application.

Example 4-18: Find the change in pressure (P_2) of a gas due to a change of temperature and volume.

Given: initial pressure $P_1 = 80$ lbs/in², initial volume $V_1 = 100$ in³, initial absolute temperature $T_1 = 500\,°F$, change in volume $V_2 = 40$ in³, change in absolute temperature $T_2 = 600\,°F$

Solution: $\dfrac{P_1 \times V_1}{T_1} = \dfrac{P_2 \times V_2}{T_2}$ and

$$P_2 = \frac{P_1 \times V_1 \times T_2}{V_2 \times T_1} = \frac{80 \times 100 \times 600}{40 \times 500}$$

$$= \frac{4{,}800{,}000}{20{,}000} = 240 \text{ lbs/in}^2$$

Chapter 5

STRENGTH OF MATERIALS

Strength of materials is a branch of mechanics dealing with the behavior of materials under stress from external forces. Familiarity with strength of materials enables one to properly select materials for use in machine parts and in members of permanent structures. (See Chapter 4, MECHANICS, for basic information.)

The materials commonly used in manufacturing and construction are metals (and their alloys), plastics, woods, concrete, asphalt mixtures, and clays. Each of these materials has properties that make it suitable for particular applications, but not for other applications. For example, ductile materials such as mild steel and aluminum are suitable for applications that brittle materials such as cast iron or concrete are not suited for.

BASIC CONCEPTS AND TERMS

The basic concepts and terms covered here are unique to the study of strength of materials. Knowledge of *working load, stress, deformation, strain,* and *modulus of elasticity* is essential for solving problems related to the strength of materials.

Working Load

The working load is the sum of all known external forces (loads) applied to a restrained solid (a machine part or a construction member). The external forces may be any one of the following four, or a combination of them:

1. *The force of a static load,* as in the case of a large water tank on poles or the weight held by a crane or an elevator.
2. *The force of momentum* caused by a speed change of a part with a certain mass, as in the case of a machine part that has reciprocating motion.
3. *The centrifugal force* of a rotating mass, as in the case of a rotating crankshaft wheel.
4. *A frictional force,* as in the case of a car braking—that is, the brake shoes on the drums of the wheels.

The external forces applied to a machine part, or member of a permanent construction, are balanced (neutralized) by the internal resisting force of the material from which the part or member is made. This is true only for external forces with a magnitude that does not exceed the given value of the resisting force of the material. In cases where the external forces are too high, that is, their given value exceeds the resisting force value, the part deforms and eventually breaks.

Stress

Stress is the effect of an external force applied upon a solid material. (See WORKING STRESS in this chapter.) To maintain equilibrium, the solid material has an internal resistance that absorbs the external force. This internal resistance is expressed in pounds per square inch (lb/in^2 or psi). Every machine part or structure member is designed to safely withstand a certain amount of stress. The position and direction of the stress determines what type of stress it is. Five types of stress (Figure 5-1) are: *tensile, compressive, shearing, bending,* and *torsional.* (Each type of stress is covered in depth later in this chapter.)

The materials from which machine parts and structure members are made have varying degrees of strength under each type of stress.

NOTE: The strength of one of the most commonly used materials, steel, can be increased through a heat-treating process called hardening. That is, steel becomes stronger after hardening and, therefore, it can withstand a greater amount of stress.

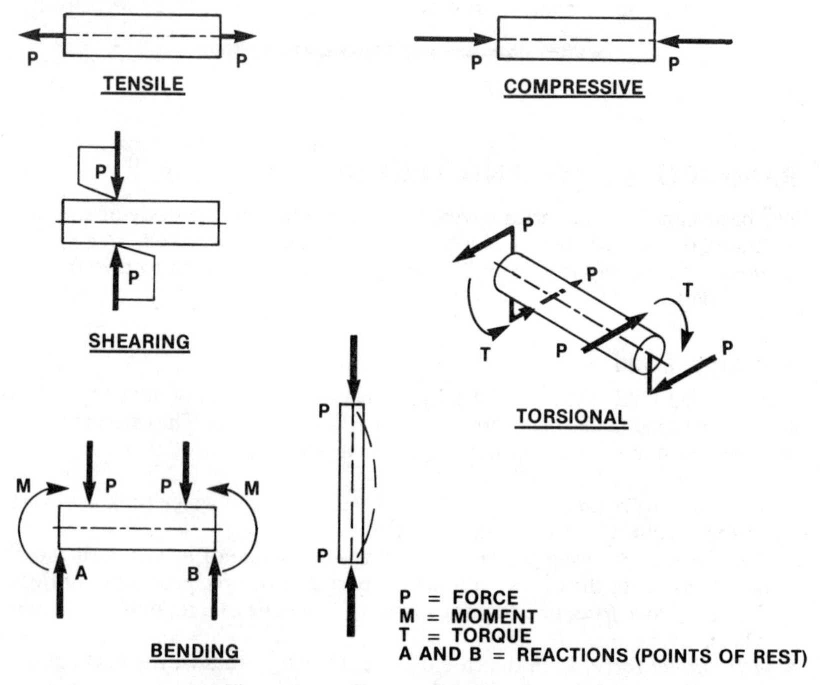

Figure 5-1. The five types of stresses.

Deformation

Deformation is a change in shape and size of a solid material, or just a change in shape of a solid material. This change is caused by excessive stress, and each type of stress causes a different deformation. For example: (1) tensile stress causes a reduction in the cross-sectional area of a solid and a lengthening of its original size; (2) compressive stress causes an increase in the cross-sectional area of a solid and a shortening of its original length; (3) shearing, bending, and torsional stress only cause a change in the shape of a solid.

Strain

Strain is the deformation per unit length of a solid under stress. The magnitude of strain is equal to the total amount of deformation divided by the original size of the solid, and is usually expressed as a percentage of linear units—for instance, percent of inch per inch (% in/in).

Modulus of Elasticity

Modulus of elasticity of a material is an index of its elasticity, or the ability of a solid material to deform when an external force is applied to it, then return to its original shape after the removal of that external force. The numerical value of this index, or constant, is unique to each material. It is related to stress and strain and is expressed in pounds per square inch (lb/in^2 or psi). Figure 5-2 shows the values of the modulus of elasticity for tensile, compressive, and shearing stresses of common materials.

MATERIAL	MODULUS OF ELASTICITY IN PSI	
	TENSION & COMPRESSION E	SHEAR E_s
Steel	30,000,000	12,000,000
Wrought iron	27,000,000	10,000,000
Malleable cast iron	24,000,000	10,000,000
Gray cast iron	15,000,000	6,000,000
Copper, cast	13,000,000	6,000,000
Copper, hard drawn	17,000,000	6,000,000
Brass (60% copper, 40% zinc), cast	13,000,000	5,000,000
Bronze (90% copper, 10% tin), cast	12,000,000	—
Aluminum and its alloys	10,000,000	4,000,000
Magnesium and its alloys	6,500,000	2,400,000
Lead, rolled	1,000,000	—
Concrete	2,000,000–3,000,000†	—
Rubber compounds	180–1,200	50–200
Epoxy cast resins, no filler	350,000	—
Polystyrene	450,000	—
Polyethylene	100,000	—
Polyvinyl-chloride	300,000	—

†For compressive stress only. (Ordinarily, concrete is not considered to have tensile strength.)

Figure 5-2. Values of modulus of elasticity of selected engineering materials.

TENSILE STRESS

Tensile stress is the stress caused by two equal forces acting on the same axial line of an object, but in opposite directions. Tensile stress tends to stretch the object. If the amount of tensile stress exceeds a certain value beyond the *proportional limit* (*elastic limit*—see TENSILE STRESS TEST) of the material, it causes the part to deform permanently. This deformation reduces the cross-sectional area of the part and correspondingly increases its length (Figure 5-3).

The magnitude of stress to which a part is subjected depends on the amount of external force placed on it, and the cross-sectional area of the part. The same external force causes greater stress to a part with a smaller cross-sectional area than to a part with a larger cross-sectional area.

The behavior of materials under stress is observed by conducting various tests under controlled conditions. The most common test is the *tensile stress test*. Information derived from this test is used to solve problems related to the strength of materials.

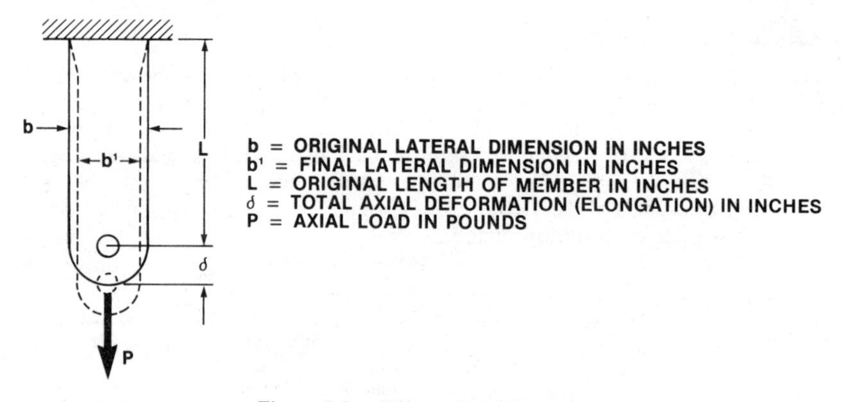

Figure 5-3. Effect of tensile stress.

Tensile Stress Test

The tensile stress test is an axial stretching test of a specimen made from the same material whose strength is tested. The specimen has a specified shape with accurate dimensions and is mounted in a universal testing machine (Figure 5-4). Then, a constantly increasing external force, or load, is applied at each end of the specimen, but in opposite directions (Figure 5-4). The specimen resists up to a certain point, deforms (elongates), and eventually breaks or fractures. According to *Hooke's Law*, stress is proportional to strain, and when the external force or load is removed, the material returns to its original condition. During the tensile stress test, the value of the load and the corresponding increase in length (elongation) of the specimen are recorded and plotted in a diagram (Figure 5-5). The various points on the diagram provide the following information:

1. *Proportional limit or elastic limit* indicates the maximum stress the material can withstand without permanent deformation. Up to this point, the strain is

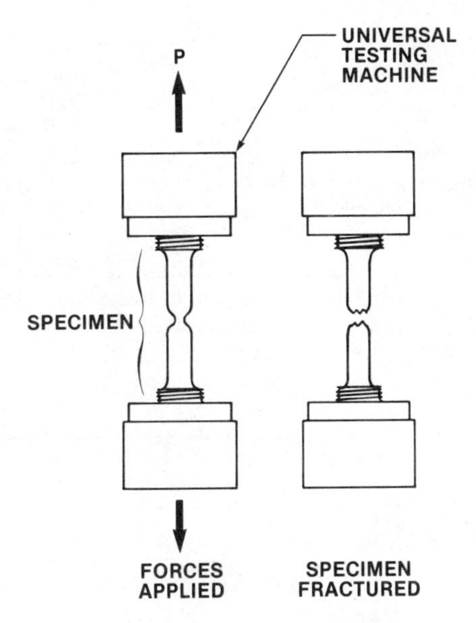

Figure 5-4. Tensile stress test.

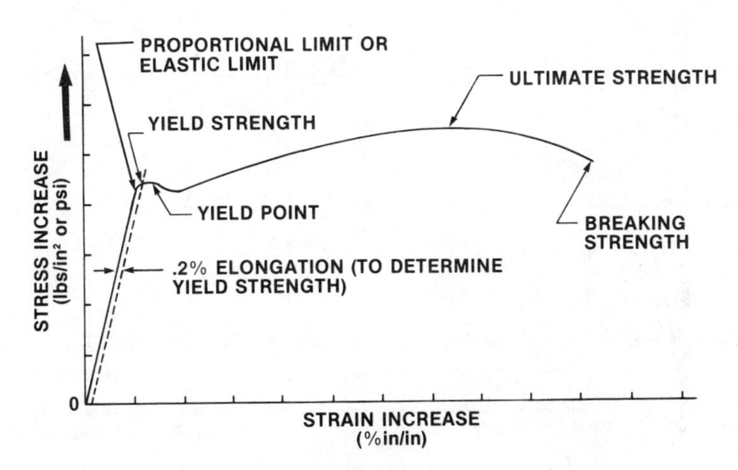

Figure 5-5. Tensile test diagram for steel.

proportional to the stress, and the specimen returns to its original shape and length when the load or external force is removed.

2. *Yield strength* indicates the amount of stress required to cause 0.2% elongation.
3. *Yield point* indicates the point at which a specific, limited elongation occurs without any increase in stress.

4. *Ultimate strength* indicates the maximum stress the material can withstand.
5. *Breaking strength* indicates the point at which the specimen breaks into two pieces.

NOTE: For some materials, the breaking strength is the same as the ultimate strength, but for other materials, the breaking strength has a lower value, as in the case of the steel specimen in Figure 5-5.

The mechanical properties of any material are determined through testing. The testing of the common materials used in industry and construction reveals that each material, under stress, exhibits certain behavior before it breaks or fractures. This behavior is shown graphically in the diagram of Figure 5-6.

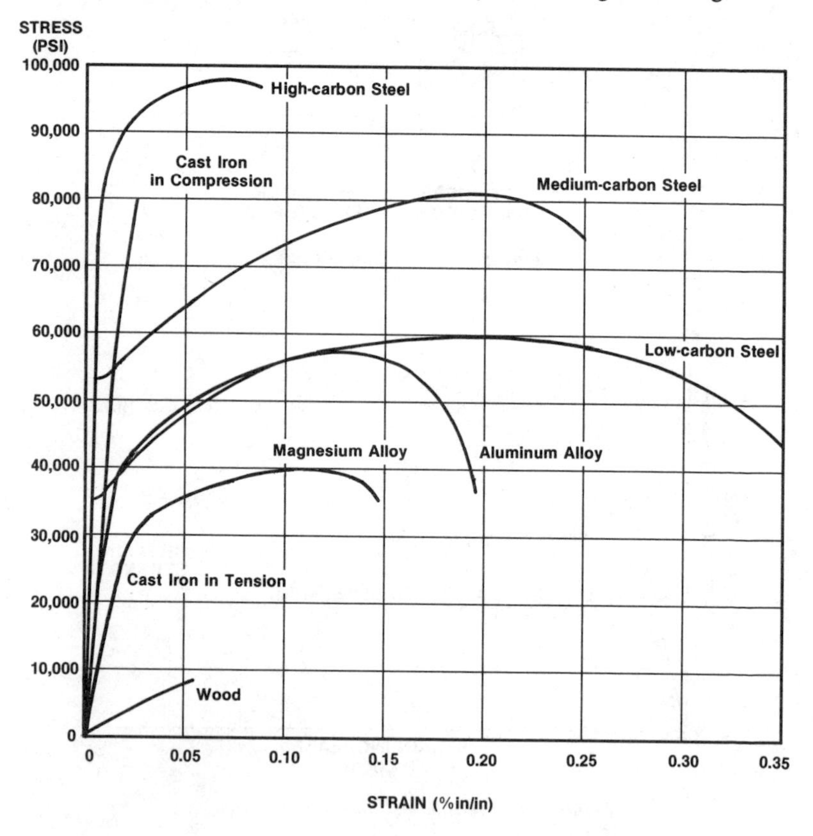

Figure 5-6. Tensile stress test diagram of selected engineering materials.

The strength of materials in other types of stress is different from their tensile strength. For example, the ultimate strength of hot-rolled carbon steel with 0.2% carbon content is 60,000 lbs/in² in tension, 90,000 lbs/in² in compression, and 45,000 lbs/in² in shear (Figure 5-7).

MATERIAL	PROPORTIONAL LIMIT (lb/in²)			ULTIMATE STRENGTH (lb/in²)			
	TENSION	COMPRESSION	SHEAR	TENSION	COMPRESSION	SHEAR	ELONGATION in 2″
Steel, 0.2% carbon,							
hot-rolled	35,000	35,000	21,000	60,000	90,000	45,000	35.0
annealed castings	30,000	30,000	18,000	60,000	*	45,000	30.0
Steel, 0.6% carbon,							
hot-rolled	60,000	60,000	36,000	100,000	*	80,000	15.0
Steel, 1.0% carbon,							
hot-rolled	80,000	80,000	48,000	135,000	*	115,000	10.0
oil-quenched	135,000	135,000	80,000	320,000	*	185,000	1.0
Nickel steel							
(3.5% nickel)							
(0.4% carbon)							
oil-quenched	160,000	160,000	96,000	285,000	*	—	5.0
Wrought iron	30,000	30,000	18,000	50,000	60,000	40,000	30.0
Gray cast iron							
ATM No. 25†	6,000	25,000	—	26,000	98,000	35,000	Negligible
Copper							
Annealed	3,200	3,200	1,900	32,000	*	—	56.0
Hard-drawn	38,000	38,000	23,000	55,000	*	—	4.0
Brass							
(60% copper)							
(40% zinc)							
Cast	20,000	—	—	50,000	—	—	20.0

*Taken as equal to the proportional limit in tension.
†92.9% iron, 3.5% carbon, 2.6% silicon, 0.6% manganese, 0.1% sulphur, 0.3% phosphorus.

Figure 5-7. Average values of strength and elongations for selected engineering materials.

Formulas Derived from the Tensile Stress Test. The tensile stress test provides information related to stress, strain, and deformation from which the following mathematical formulas are derived:

Stress: $\sigma = \frac{P}{A} = \text{lb/in}^2$

Deformation: $\delta = \frac{P \times L}{A \times E} = \text{in}$
(elongation)

Strain: $\varepsilon = \frac{\sigma}{E} = \text{in/in}$ or

$\varepsilon = \frac{\delta}{L} = \text{in/in}$

σ = perpendicular force per unit area in pounds per square inch (lbs/in² or psi)
P = perpendicular force in pounds (lbs)
A = cross-sectional area on which the perpendicular force (load) is applied in square inches (in²)
ε = deformation (special elongation) per unit length in inches per inch (in/in)
E = constant, called *Young's Modulus of Elasticity* in pounds per square inch (lb/in² or psi) applied to tensile and compressive stress
δ = deformation (elongation) in inches
L = original length of the specimen in inches

These formulas and the other information derived from the tensile stress test are used to solve problems related to the strength of materials. See Example 5-1 for an application of these formulas.

Example 5-1: A square steel rod is 12″ long with a cross-sectional area of 0.25 square inches. It is subjected to tension by 10,000 pounds of force. Find the stress, deformation, and strain to the rod.

Given: $L = 12''$, $A = 0.25\ \text{in}^2$, $P = 10{,}000$ lbs
Numerical value of modulus of elasticity of steel is:
$E = 30{,}000{,}000\ \text{lb/in}^2$ (from table in Figure 5-2)

Solution:

Stress: $\sigma = \dfrac{P}{A} = \dfrac{10{,}000}{0.25} = 40{,}000\ \text{lb/in}^2$

Deformation: $\delta = \dfrac{P \times L}{A \times E} = \dfrac{10{,}000 \times 12}{0.25 \times 30{,}000{,}000} = \dfrac{120{,}000}{7{,}500{,}000} = 0.016''$
(elongation)

Strain: $\varepsilon = \dfrac{\sigma}{E} = \dfrac{40{,}000}{30{,}000{,}000} = 0.00133\ \text{in/in}$ or

$$\varepsilon = \frac{\delta}{L} = \frac{0.016}{12} = 0.00133\ \text{in/in}$$

Each part of a machine or each structure member has a certain strength. This strength depends on the type of material from which the part or member is made. The strength also depends on the smallest cross-sectional area of the part or structure member. When the cross-sectional area under stress is not uniform throughout, the smallest cross-sectional area of the part or member is used to determine the strength. See Example 5-2.

Example 5-2: A round steel bar, 0.750″ in diameter, has a 0.650″ diameter groove equidistant from both ends of the bar. It is subjected to a stress of 10,000 pounds of force. Find the stress which this force causes to the section areas corresponding to the large diameter and small diameter of the steel bar.

Given: large diameter $d_1 = 0.750''$, small diameter $d_2 = 0.650''$,
force $P = 10{,}000$ lbs

The formula for finding the area of a circle is: $A = \dfrac{\pi \times d^2}{4}$ or $0.785\ d^2$

therefore: $A_1 = (0.785)(0.750)^2 = 0.442\ \text{in}^2$ and
$A_2 = (0.785)(0.650) = 0.332\ \text{in}^2$

Solution: $\sigma_1 = \dfrac{P}{A_1} = \dfrac{10{,}000}{.442} = 22{,}624\ \text{lb/in}^2$

$$\sigma_2 = \frac{P}{A_2} = \frac{10{,}000}{.332} = 30{,}120\ \text{lb/in}^2$$

NOTE: The same force (10,000 lbs) causes much greater stress to the smallest of the two cross-sectional areas.

Working Stress

Working stress is the stress the material of a machine part or structure member is expected to bear. (See STRESS in this chapter.) Working stress depends on the sum of all known applied external forces and the cross-sectional area of the part or member. To avoid any permanent deformation, the working stress must be below the elastic limit of the material. Avoiding permanent deformation is assured by using a *factor of safety* in connection with the ultimate strength or the yield strength of materials.

NOTE: When determining the working stress for shearing, the factor of safety is used in connection with the ultimate strength or yield strength of the materials in shearing, not tension, because the strength of materials in shearing is lower than in tension and compression. The ultimate strength or yield strength of materials in shearing is also used to determine the working stress of materials subjected to bending and torsion.

Factor of Safety. Factor of safety is a number denoting the quotient of either the ultimate strength or the yield strength of a material, and the expected working stress of a part of a machine or structure made from this material. Factor of safety indicates how many times smaller the working stress must be, compared to the ultimate strength or yield strength of the material.

The numerical value (f) of the factor of safety ranges from 2 to 10. In every application, the choice of a factor of safety is determined by considering the type of stress and the desired degree of safety that a part of a machine or structure must provide. This relationship is shown in the following formulas:

$$\text{Working stress} = \frac{\text{Ultimate strength}}{\text{Factor of safety}} \quad \text{or} \quad \frac{\text{Yield strength}}{\text{Factor of safety}}$$

$$\text{Factor of safety} = \frac{\text{Ultimate strength}}{\text{Working stress}} \quad \text{or} \quad \frac{\text{Yield strength}}{\text{Working stress}}$$

NOTE: The same factor of safety provides a greater degree of safety if it is used in connection with the yield strength than with the ultimate strength. This higher degree of safety is achieved by increasing the cross-sectional area of the part. See Example 5-3. The same formulas and the other information presented earlier in this chapter are used to solve problems related to compressive stress, and with slight modifications, problems related to the other simple stresses.

Example 5-3: A square steel rod is to be used in a structure which is subjected to a simple tension stress by 80,000 pounds force. The ultimate strength of the material from which the rod is made is 60,000 pounds per square inch, and the yield strength is 30,000 pounds per square inch. Find the cross-sectional area of this rod by using the number 3 as factor of safety.

Given: Force = 80,000 lbs, ultimate strength = 60,000 lb/in^2,
yield strength = 30,000 lb/in^2, factor of safety = 3

Solution: 1. $\text{Working stress} = \dfrac{\text{Ultimate strength}}{\text{Factor of safety}} = \dfrac{60{,}000}{3} = 20{,}000\ \text{lb/in}^2$

$$\text{Area} = \frac{\text{Force}}{\text{Working stress}} = \frac{80{,}000}{20{,}000} = 4\ \text{in}^2$$

2. $\text{Working stress} = \dfrac{\text{Yield strength}}{\text{Factor of safety}} = \dfrac{30{,}000}{3} = 10{,}000\ \text{lb/in}^2$

$$\text{Area} = \frac{\text{Force}}{\text{Working stress}} = \frac{80{,}000}{10{,}000} = 8\ \text{in}^2$$

NOTE: In the second case, the greater safety is achieved by using a rod with a cross-sectional area that is greater than the first case.

COMPRESSIVE STRESS

Compressive stress is the stress caused by two equal forces acting on the same axial line of a solid, but in opposite directions. Compressive stress tends to squeeze the solid (part). The deformation caused by compressive forces consists of an increase in the cross-sectional area and a decrease in the original length of the solid (Figure 5-8).

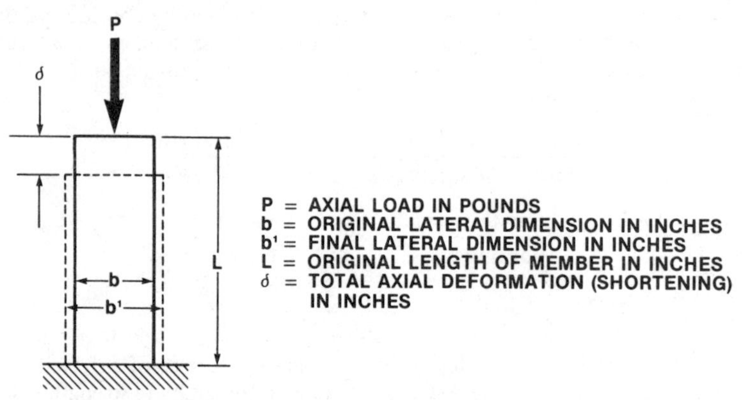

Figure 5-8. Effect of compressive stress.

The strength of materials in compression is determined by a compressive test conducted under controlled conditions. The compressive strength of ductile materials is almost equal to their strength in tension. Hooke's Law (see TENSILE STRESS TEST) is applied to compression provided that the length of the part under compressive stress is not greater than eight times its diameter, or eight times its smallest dimension (side). If the length of the part subjected to a compressive stress is greater than eight times its smallest dimension, the part fails and bends before reaching the ultimate strength of the material. For this reason, in problems related to compressive stress, the length of the part must be considered. See Example 5-4.

Example 5-4: A compressive force of 120,000 pounds (the permissible load) is applied to a rectangular cast-iron block, 1½″ × 4″ × 10″. Find the stress that this force causes to the cast-iron block, and the numerical factor of safety if the ultimate strength of the material is 60,000 lbs/in².

Given: P = 120,000 lbs, A = 6 in² (1.5 × 4 = 6 in²), ultimate strength of material = 60,000 lb/in²

Solution: $\text{Stress—}\sigma = \frac{P}{A} = \frac{120{,}000}{6} = 20{,}000 \text{ lb/in}^2$

$$\text{Factor of safety—f} = \frac{\text{Ultimate strength}}{\text{Working stress}} = \frac{60{,}000}{20{,}000} = 3$$

SHEARING STRESS

Shearing stress is the stress caused by two equal and parallel forces acting upon a solid in opposite directions. Shearing stress causes one side of the solid to "slide" in relation to the other side. If the amount of shearing stress exceeds a certain value, the part or member deforms permanently. This deformation results in a change in the shape of the part and in the creation of an angle (γ) (Figure 5-9).

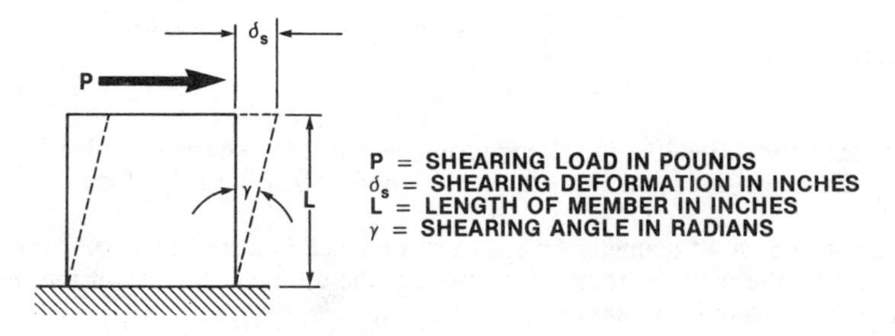

Figure 5-9. Effect of shearing stress.

The total deformation in shearing occurs over the length L and is represented by the symbol (δ_s). The deformation per unit of length is expressed in radians and is represented by angle (γ)—the *shearing angle*. The shearing angle is for shearing, as strain is for tension and compression.

Shearing stress placed on the cross-sectional area of a solid is parallel to the force, not perpendicular, as in the case of tensile stress and compressive stress. See Figure 5-10 for a comparison between perpendicular stress and parallel stress.

The strength of materials in shearing stress is less than it is in tensile and compressive stresses. However, Hooke's Law is also applied to shearing. The same formulas, with minor adjustments, are used to solve problems related to shearing:

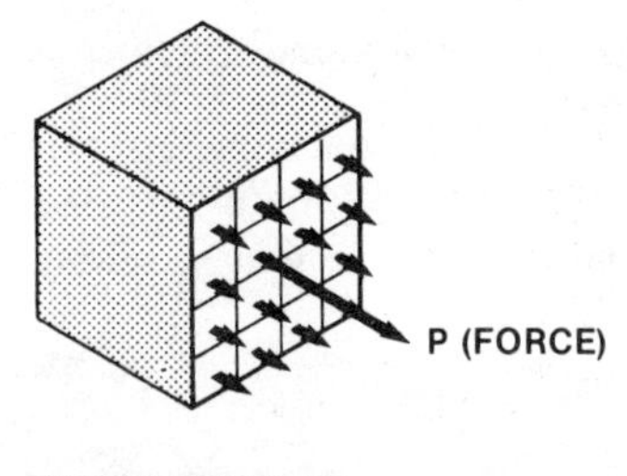

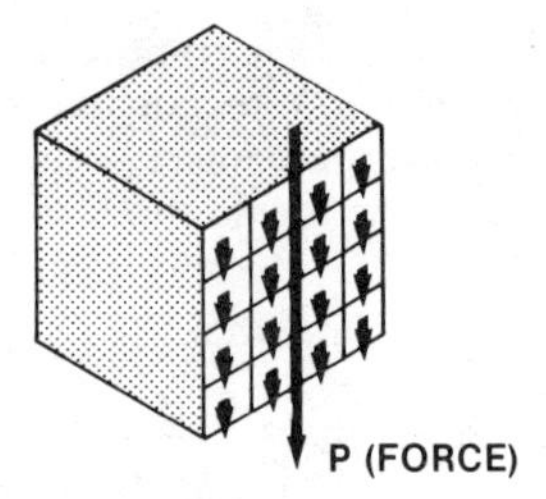

PARALLEL
(SHEARING STRESS)

Figure 5-10. Comparison of perpendicular and parallel stresses.

Shearing stress: $\sigma_s = \dfrac{P}{A_s}$

Shearing angle: $\gamma = \dfrac{\delta_s}{L}$

Shearing modulus of elasticity $E_s = \dfrac{\sigma_s}{\gamma}$

σ_s = shearing stress in pounds per square inch (lbs/in² or psi)

P = force in pounds (lbs)

A_s = area in inches squared (in^2)

γ = shearing angle in radians

E_s = shearing modulus of elasticity in pounds per square inch (lbs/in^2)

NOTE: The symbols in these formulas are followed by the subscript s to indicate shearing.

One of the difficulties in solving problems related to shearing is identifying the solid's cross-sectional area subjected to the shearing stress. See Example 5-5.

Example 5-5: A 1″ diameter hole is to be punched in a steel plate 0.5″ thick. Determine the required force of the press if the ultimate strength of the steel plate in shearing is 40,000 lb/in^2.

Given: Diameter of hole d = 1″, thickness of plate h = 0.5″, ultimate strength of the material σ_s = 40,000 lb/in^2

The shearing cross-sectional area is equal to the perimeter of the hole times the thickness of the plate: $A_s = \pi dh = 3.14 \times 1 \times 0.5 = 1.57\ in^2$

Solution: $\sigma_s = \dfrac{P}{A_s}$ and $P = \sigma_s \times A_s$

$P = 40{,}000 \times 1.57 = 62{,}800$ lbs

BENDING STRESS

Bending stress placed on a solid resting on one or more points is caused by a load or force acting perpendicular to the solid's horizontal axis. This load or force bends the solid. The deformation caused by excessive bending stress changes the shape of the solid and creates a deflection identified as δ (Figure 5-11).

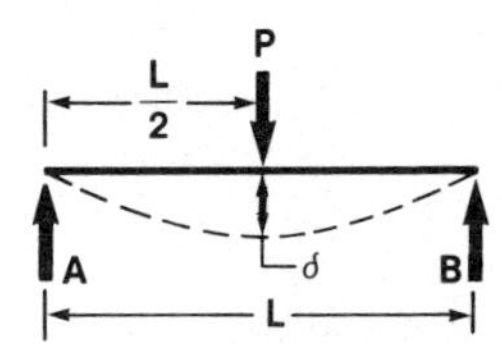

L = TOTAL LENGTH OF BEAM
P = PERPENDICULAR FORCE
δ = TOTAL DEFORMATION (DEFLECTION)
B = REACTION OR REACTING FORCE
A = REACTION OR REACTING FORCE

NOTE: FOR EQUILIBRIUM, THE ALGEBRAIC SUM OF ACTING AND REACTING FORCES MUST BE ZERO.

Figure 5-11. Effect of bending stress on a beam.

NOTE: Bending is associated with beams and columns. The bending stress developed on a beam by the applied external forces depends on the maximum bending moment (product of the applied load times the distance from the fulcrum) and properties of the cross-sectional area of the beam.

Bending stress is found by using the *flexure formula*:

$$\sigma = \frac{Mc}{I} \quad \text{or} \quad \sigma = \frac{M}{Z}$$

σ = maximum stress on the cross-sectional area in pounds per square inch (lbs/in² or psi)
M = maximum bending moment in inch pounds (lb-in)
c = distance from the neutral axis to the farthest point in the cross section in inches
I = moment of inertia in inches to the 4th power (in^4)

Z = section modulus in cubic inches (in^3); $Z = \frac{I}{c}$

For an application of the flexure formula, see Example 5-6.

Example 5-6: A solid shaft, 1″ in diameter, is subjected to a maximum bending moment of 1,200 lb in. Find the bending stress.

Given: diameter d = 1″, maximum bending moment M = 1,200 lb in, distance from neutral axis to the cross-sectional area c = 0.5″, moment of inertia $I = \frac{\pi d^4}{64}$ (obtained from Figure 5-12)

$$I = \frac{3.14(1)^4}{64} = \frac{3.14}{64} = 0.049 \text{ in}^4$$

$$\text{Solution: } \sigma = \frac{Mc}{I} = \frac{1{,}200(0.5)}{0.049} = \frac{600}{0.049} = 12{,}245 \text{ lb/in}^2$$

NOTE: The moment of inertia I, and the section modulus Z, as well as other properties of solids with cross-sectional areas having regular shapes, can be found by using the formulas in Figure 5-12.

REGULAR-SHAPED SECTION A	AREA OF SECTION A	DISTANCE FROM NEUTRAL AXIS c	MOMENT OF INERTIA I	SECTION MODULUS $Z = \frac{I}{c}$	RADIUS OF GYRATION $k = \sqrt{\frac{I}{A}}$
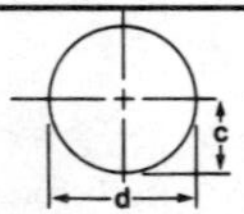 c, d	$\frac{\pi d^2}{4} = 0.785\ d^2$	$\frac{d}{2}$	$\frac{\pi d^4}{64} = .049\ d^4$	$\frac{\pi d^3}{32} = .098 d^3$	$\frac{d}{4}$
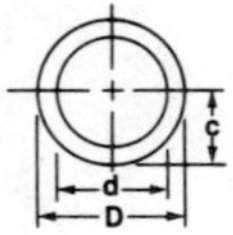 c, d, D	$\frac{\pi(D^2 - d^2)}{4} =$ $.785(D^2 - d^2)$	$\frac{D}{2}$	$\frac{\pi(D^4 - d^4)}{64} =$ $.049(D^4 - d^4)$	$\frac{\pi(D^4 - d^4)}{32D} =$ $.098\left(\frac{D^4 - d^4}{D}\right)$	$\frac{\sqrt{D^2 + d^2}}{4}$
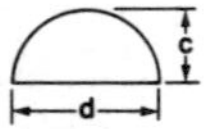 c, d	$\frac{\pi d^2}{8} = .393 d^2$	$\frac{(3\pi - 4)d}{6\pi} =$ $.288d$	$\frac{(9\pi^2 - 64)d^4}{1,152\pi} =$ $.007d^4$	$\frac{(9\pi^2 - 64)d^3}{192(3\pi - 4)} =$ $.024d^3$	$\frac{\sqrt{(9\pi^2 - 64)d^2}}{12\pi} =$ $.132d$
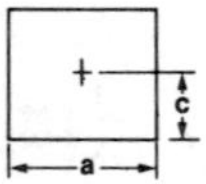 c, a	a^2	$\frac{1}{2} a = \frac{a}{2}$	$\frac{a^4}{12}$	$\frac{a^3}{6}$	$\frac{a}{\sqrt{12}} = .289a$
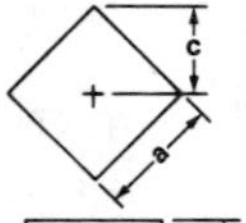 c, a	a^2	$\frac{a}{\sqrt{2}} = \frac{a}{1.414}$	$\frac{a^4}{12}$	$\frac{a^3}{6\sqrt{2}} = .118a^3$	$\frac{a}{\sqrt{12}} = .289a$
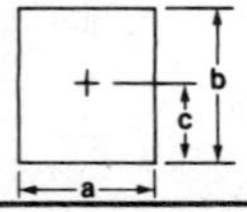 b, c, a	ba	$\frac{1}{2} a = \frac{a}{2}$	$\frac{ba^3}{12}$	$\frac{ba^2}{6}$	$\frac{a}{\sqrt{12}} = .289a$

Figure 5-12. Formulas for the properties of solids whose cross-sectional areas are regular shapes.

In problems related to bending, the maximum bending moment and the maximum deflection of a beam caused by the application of known external forces can be found by using various formulas.

NOTE: These formulas and the properties of cross-sectional areas of standard structural beams and columns are listed in various publications which are published by technical societies such as the American Institute of Steel Construction.

Each formula is applicable only for a certain type of beam and load. For example, the maximum bending moment (M) and the maximum deflection (σ) of a beam like the one shown in Figure 5-11, are found by using the following formulas:

Maximum bending moment $M = \frac{PL}{4}$

Maximum deflection $\sigma = \frac{PL^3}{48\ EL}$

M = moment (in lb)

E = modulus of elasticity (lb/in^2)
P = load (lb)
L = length in inches

Bending of Slender Columns

A slender column is a column that has a height greater than eight times its diameter, or eight times its smallest side if it is a non-cylindrical column. The bending of a slender column is caused by two equal and opposite compressive forces placed at both ends (Figure 5-13). The deflection occurs before the bending stress on the column reaches the ultimate strength of the column material.

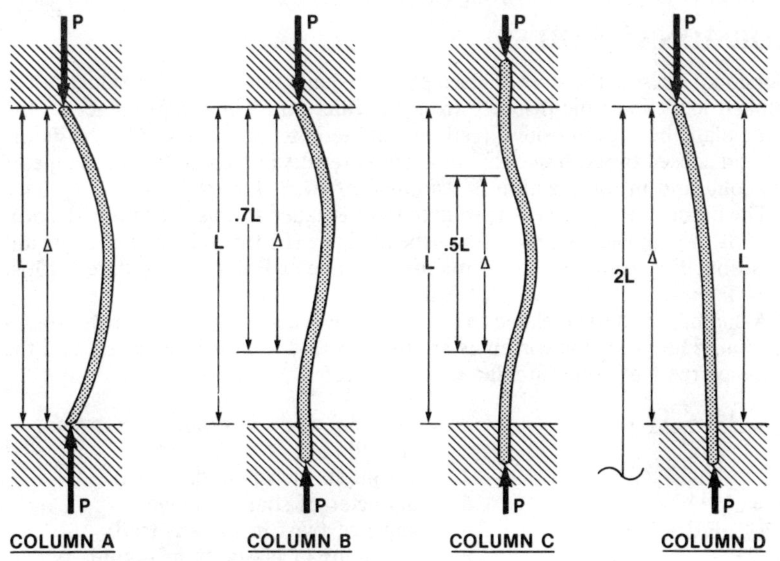

Figure 5-13. Typical examples of column-end restraints.

The strength of a column depends not only on the column material and cross-sectional shape, but also on the way in which its ends are restrained. Figure 5-13 shows typical examples of column-end restraints. The average unit stress is found by using these respective formulas:

Column A: $\frac{P}{A} = \frac{\pi^2 E}{(L/k)^2}$

Column B: $\frac{P}{A} = \frac{\pi^2 E}{(.7L/k)^2} = \frac{2.05\pi^2 E}{(L/k)^2}$

Column C: $\frac{P}{A} = \frac{\pi^2 E}{(.5L/k)^2} = \frac{4\pi^2 E}{(L/k)^2}$

Column D: $\frac{P}{A} = \frac{\pi^2 E}{(2L/k)^2} = \frac{\frac{1}{4}\pi^2 E}{(L/k)^2}$

$\frac{P}{A}$ = average unit stress on the cross-sectional area supporting buckling load P in pounds per square inch (lb/in^2)
E = modulus of elasticity of the material in pounds per square inch (lb/in^2)
L/k = slenderness ratio:
L = effective length of column in inches
k = least radius of gyration of cross-sectional area in inches

NOTE: In each formula, the *effective length* is used instead of the actual length of the column (Figure 5-13). Effective length is the distance between adjacent points of maximum deflection (Δ). The effective length for Column A is equal to the actual length (L); for Column B, the length is 0.7L; for Column C, the length is 0.5L; and for Column D, the length is 2L. Column B is two times as strong as Column A; Column C is four times as strong as Column A; and Column D is ¼ times as strong as Column A.

TORSIONAL STRESS

Torsional stress is the stress caused by two equal moments (the product of the applied load times the distance from the fulcrum). The moments act on the same plane but in opposite directions, and tend to twist the solid. The deformation caused by excessive torsional stress results in a change in the shape of the solid and in the creation of an *angle of twist* (Figure 5-14).

The moment that causes the solid to twist is called *torque*. Torque is the product of the applied force (P) times the distance (L) from the center of its application. For example, the torque on the pipe at Point A in Figure 5-15 is: $T = P \times L = 120 \times 16 = 1{,}920$ lb in.

A common problem related to torsion involves a shaft used to transfer rotary motion. The following formulas are used to find the torsional stress and the angle of twist of solid circular shafts subjected to torsion:

$$\sigma_s = \frac{16 \times T}{\pi d^3}$$

$$\theta = \frac{32TL}{\pi E_s d^4}$$

σ_s = torsional stress in pounds per square inch (lb/in^2)
T = torque in inch pounds (lb in)
d = diameter of shaft in inches
θ = angle of twist in radians (rad)
E_s = modulus of elasticity in pounds per square inch (lb/in^2)

See Examples 5-7 and 5-8 for applications of these formulas.

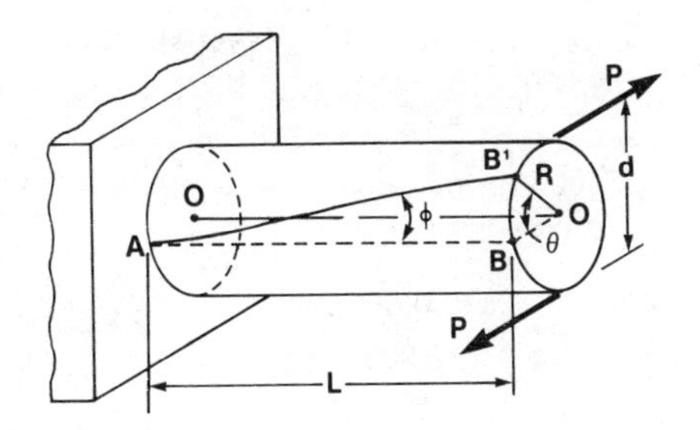

P = FORCE
d = DIAMETER
OO = SHAFT AXIS
AB = ORIGINAL POSITION OF OUTER FIBER
AB' = ASSUMED POSITION OF OUTER FIBER DUE TO TORQUE Pd
BB' = TOTAL DEFORMATION (TWIST) IN LENGTH L
R = RADIUS OF SHAFT
θ = ANGLE OF TWIST
ϕ = UNIT DEFORMATION
L = LENGTH OF SHAFT FROM FIXED POINT

NOTE: THE ARC BB', WHICH REPRESENTS THE TOTAL DEFORMATION, IS EQUAL TO THE RADIUS (R) TIMES ANGLE OF TWIST (θ) EXPRESSED IN RADIANS.

Figure 5-14. Effect of torsional stress.

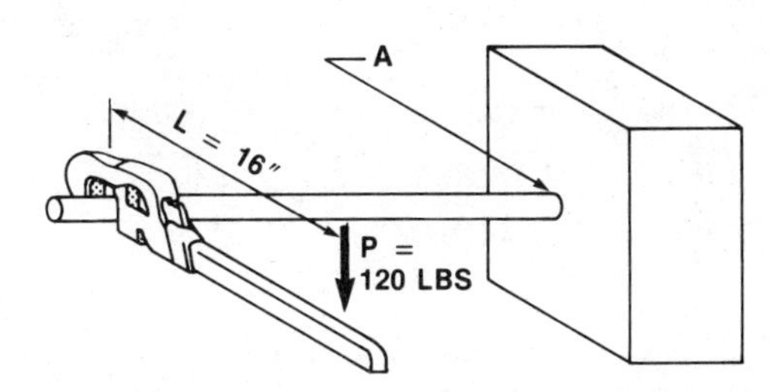

Figure 5-15. Torque when twisting a pipe with a pipe wrench.

Example 5-7: A solid circular shaft, 2" in diameter, is subjected to torsion by 10,000 in lb torque. Find the maximum torsional stress of this shaft.

Given: diameter of shaft d = 2", torque T = 10,000 in lb

Solution: $\sigma_s = \dfrac{16T}{\pi d^3} = \dfrac{16(10,000)}{3.14(2^3)} = \dfrac{160,000}{3.14(8)} = \dfrac{160,000}{25.12} = 6,370 \text{ lb/in}^2$

Example 5-8: A solid steel circular shaft, 2″ in diameter and 48″ long, is subjected to torsion by 20,000 in lb of torque. Find the angle of twist.
Given: diameter of shaft d = 2″, length of shaft L = 48″,
torque T = 20,000 in lb, $\pi = 3.14$, modulus of rigidity G = 12,000,000 lb/in²

Solution: $$\theta = \frac{32TL}{\pi E_s d^4} = \frac{32(20{,}000)(48)}{3.14(12{,}000{,}000)(2)^4} = \frac{32(960{,}000)}{37{,}680{,}000(16)}$$

$$= \frac{30{,}720{,}000}{602{,}880{,}000} = .051 \text{ radian}$$

NOTE: Radians can be converted into degrees by multiplying them by 57.3°:

$$0.051(57.3) = 2.92°$$

The diameter of the shaft can be found by using a formula derived from the torsional stress formulas:

$$\sigma_s = \frac{16T}{\pi d^3} \quad \text{therefore} \quad d = \sqrt[3]{\frac{16T}{\pi \sigma_s}}$$

For a practical application, see Example 5-9.

Example 5-9: A solid shaft made from high carbon steel is to be subjected to a torque of 40,000 in lb. Find the shaft diameter for a maximum working torsional stress of 12,000 lb/in².
Given: torque T = 40,000 in lb, stress σ_s = 12,000 lb/in²

Solution: $$d = \sqrt[3]{\frac{16T}{\pi \sigma_s}} = \sqrt[3]{\frac{16(40{,}000)}{3.14(12{,}000)}} = \sqrt[3]{\frac{640}{37.68}} = \sqrt[3]{16.985} = 2.57''$$

Chapter 6

MACHINE ELEMENTS

Machine elements are components (parts) or mechanical structures. They are manufactured from various materials and are usually produced in standard shapes and dimensions.

This chapter contains basic information related to machine elements used in permanent connections, non-permanent connections, and mechanical driving systems.

PERMANENT AND NON-PERMANENT CONNECTIONS (JOINTS)

The machine elements used in permanent connections (joints) are called *rivets*. The machine elements used in non-permanent connections are called *threaded fasteners*. In permanent joints, the joined parts cannot be disconnected without destroying the rivets. In non-permanent joints, the joined parts can be disconnected by unthreading the fasteners.

Rivets

Rivets are cylindrical metal pins that have a preformed head on one end (Figure 6-1). They are used for joining plates, structural steel components, sheet metal, and other flat parts. Riveting is accomplished by using rivets with a diameter that is slightly smaller than the diameter of the holes in the parts to be connected. Using rivets with a smaller diameter facilitates their entrance into the holes. The rivet length is greater than the combined length of the parts. This provides a sufficient amount of metal for the formation of the second head (Figure 6-2).

Figure 6-1. Typical rivets and rivet nomenclature.

Rivets are available in a variety of sizes and types of heads (Figure 6-3). The diameters of small rivets range in size from $^1/_{16}$" to $^7/_{16}$". Small rivets are used to connect thin metals or other thin materials. The formation of the second head is accomplished by applying force to a cold rivet. The diameters of large

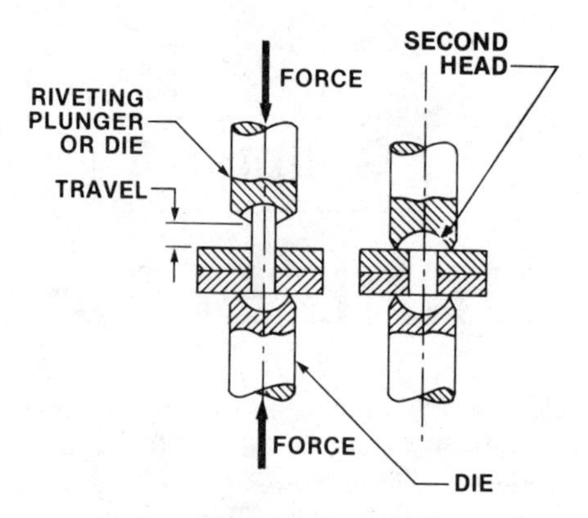

Figure 6-2. Formation of the rivet's second head.

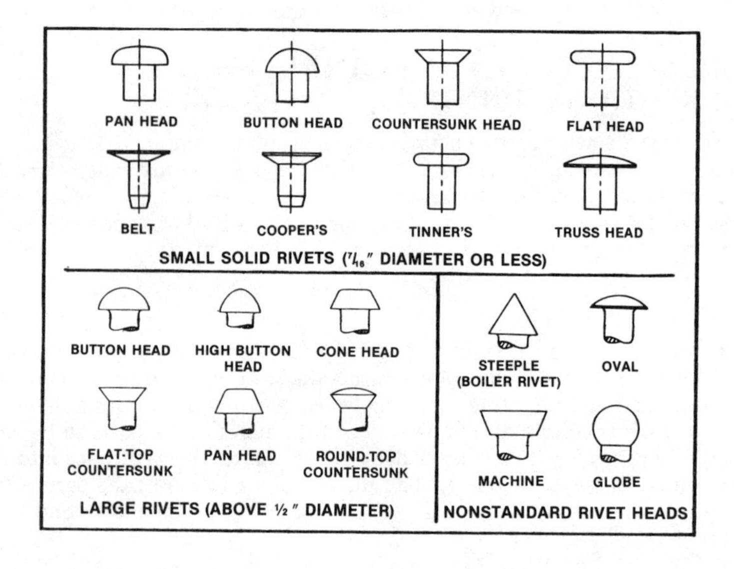

Figure 6-3. Types of rivet heads. (Machine Design)

rivets range in size from 1/2" to 1 1/2" in increments of 1/16" or 1/8". The second head of large rivets can only be formed by applying force to the rivet after it has been sufficiently heated (red hot).

Riveted Joints. The two types of riveted joints are the *lap* and *butt* joints (Figure 6-4). For lap joints, the members (plates) to be joined overlap each other; for butt joints, the plates are in the same plane and are held together by one or two butt straps placed over the seam. Both the lap joint and the butt joint are

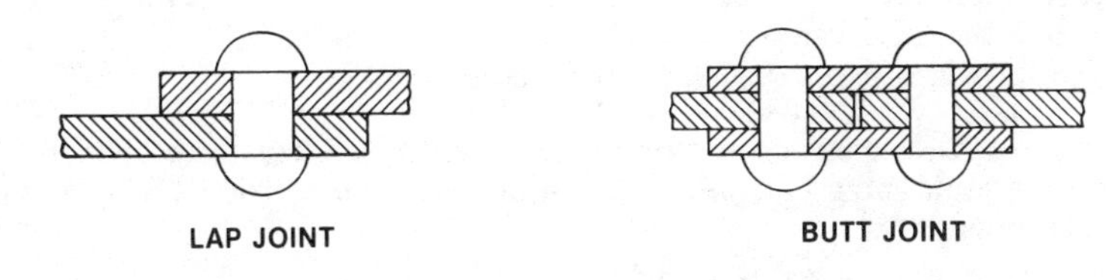

Figure 6-4. Riveted lap and butt joints.

identified by the number of rows of rivets that they have. For instance, a lap joint with a single row of rivets is called a *single-riveted lap joint*; a lap joint with two rows of rivets is called a *double-riveted lap joint*.

All the rivets used in a given riveted joint should have the same diameter to equally share the applied load to the joint. The rivet diameter depends on the load that the joint is expected to carry, and the thickness of the metals to be joined. Its diameter should be larger than the thickness of the thickest plate, but not more than three times the thickness of the thinnest plate.

CAUTION: If the diameter of the rivet is too large, the pressure required to drive the rivet may damage the thinnest plate. If the diameter is too small, the head of the rivet may pop. As a general rule, the diameter (d) of the rivets ranges from 1.2 times to 1.4 times the thickness of the thickest plate (t); that is,

$$d = 1.2t \quad \text{to} \quad d = 1.4t$$

Also, rivets having the same strength as the material being joined should be used, and the rivets should be evenly spaced.

As a general guideline, the minimum distance between the centers of the rivet holes should be at least three times their diameter, and the distance from the edge of the plates should be at least one-and-one-half times their diameter.

NOTE: Rivet manufacturers provide tables with recommendations regarding the dimensions for various types of riveted joints.

In riveted joints, the shearing stress (covered in Chapter 5) caused by an external force on a rivet depends on the magnitude of that force, the cross-sectional area of the rivet, and the number of shearing planes (Figure 6-5). The formula for finding the shearing stress of rivets is:

$$\sigma_s = \frac{P}{A_s n}$$

σ_s = unit shearing stress in pounds per square inch (lb/in^2)
P = load (force) in pounds (lbs)
A_s = shearing area of each rivet in square inches (in^2)
n = number of shearing planes

According to this formula, the stress caused to a rivet in a lap joint is twice as much as the stress caused to a rivet in a butt joint because in the butt joint, the rivet provides two shearing planes. See Example 6-1.

Example 6-1: Determine the number of one-half inch (0.5″) diameter rivets needed in a lap joint and a butt joint if the joint is to be used for carrying a load of 3,000 pounds.

Given: Maximum allowable shearing stress of the rivet material $\sigma_s = 8{,}000$ lbs/in², load $P = 3{,}000$ lbs, diameter $d = 0.5''$
$A_s = 0.785d^2$

Solution:

$$A_s = 0.785d^2 = 0.785(0.5)^2 = 0.785(0.25) = 0.2 \text{ in}^2$$

Lap joint (n = 1): $\sigma_s = \frac{P}{A_s n} = \frac{3{,}000}{0.2(1)} = \frac{3{,}000}{0.2} = 15{,}000 \text{ lbs/in}^2$

Butt joint (n = 2): $\sigma_s = \frac{P}{A_s n} = \frac{3{,}000}{0.2(2)} = \frac{3{,}000}{0.4} = 7{,}500 \text{ lbs/in}^2$

NOTE: The shearing stress in the lap joint exceeds the maximum allowable stress of 8,000 lbs/in². This means that in the lap joint, two rivets are needed while in the butt joint, only one rivet is needed to carry the same load safely.

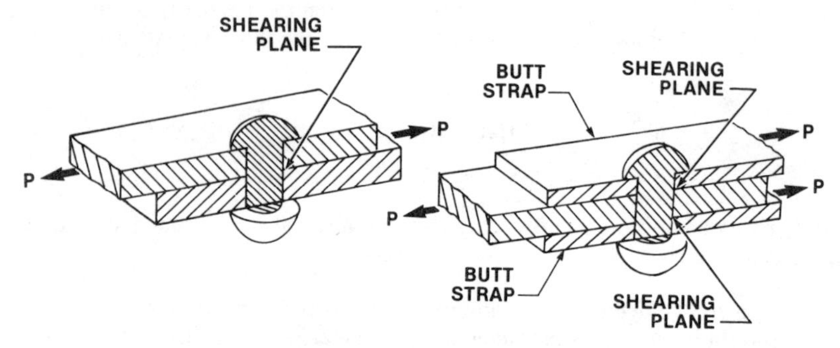

Figure 6-5. Shearing planes in riveted joints.

Threaded Fasteners

Threaded fasteners, or screw fasteners, are cylindrical rods with a head on one end and threads on the other end (Figure 6-6). They are used to connect parts for non-permanent joints—that is, joints with parts that can be assembled and disassembled without destroying the fastener. Threaded fasteners are available in standard sizes and in standard thread design forms. (See Chapter 3, STANDARD SCREW THREADS FOR FASTENERS, for detailed information.)

The strength of a threaded fastener depends on the material from which it is made and the size of the cross-sectional area of its minor diameter. In the various applications, threaded fasteners are subjected to tensile, shearing, and torsional stresses (See Chapter 5). Among these stresses, torsional stress, which results from excessive tightening torque, is the most frequent cause of breakage.

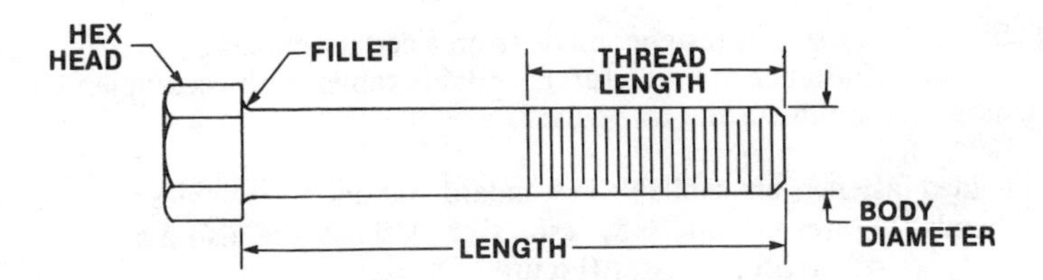

Figure 6-6. Typical threaded fastener.

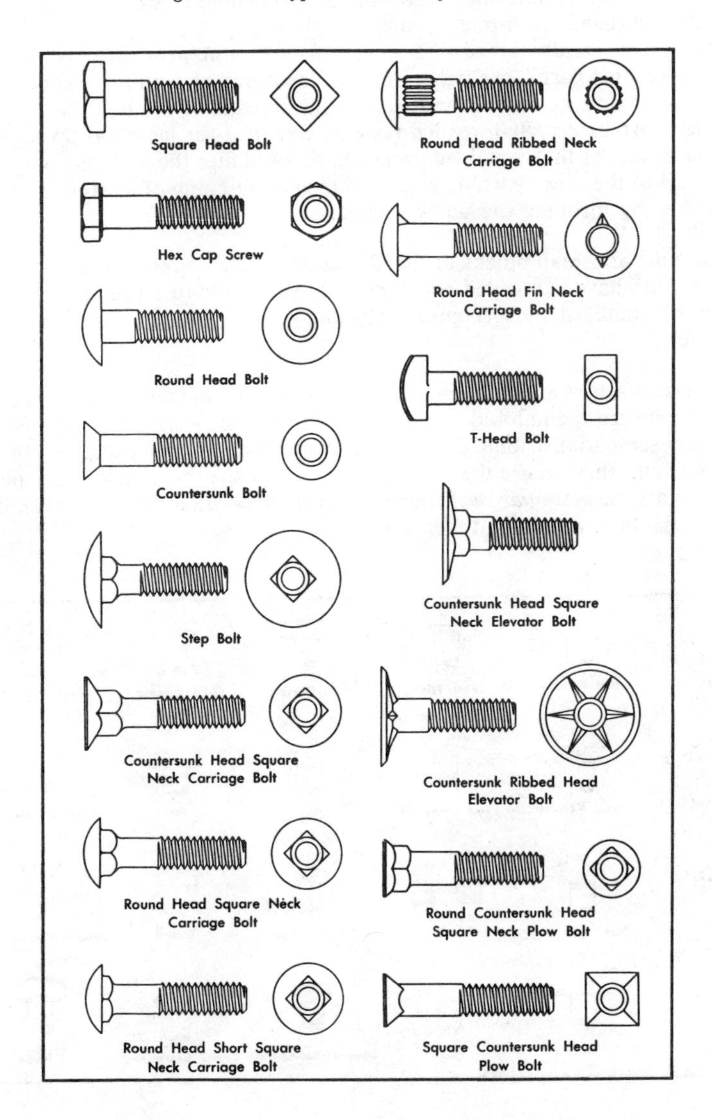

Figure 6-7. Typical standard bolt and screw styles. (Machine Design)

NOTE: For every size of fastener made from a certain material, there is a maximum torque. Fastener manufacturers provide tables with recommendations regarding maximum torque to be used.

Threaded fasteners are available in standard bolt and screw styles with a variety of heads (Figures 6-7 and 6-8, respectively). They are also available with various types of driving recesses (Figure 6-9).

NOTE: The terms *bolt* and *screw* are used to denote threaded fasteners and are differentiated from one another as follows:

Bolt: An externally-threaded fastener designed for insertion through the holes of assembled parts. The tightening and loosening of a bolt is accomplished by torquing a nut which screws onto the threads of the bolt.

Screw: An externally-threaded fastener designed for insertion through the hole of one of the assembled parts and screwed into the preformed internal thread of the other part. The tightening and loosening of a screw is accomplished by torquing the screw's head.

Nuts. Nuts are small blocks of metal that are usually square or hexagonal in shape. They have a threaded hole designed to accomodate a bolt thread. Many styles of standard nuts (Figure 6-10) and specialty nuts (Figure 6-11) are available.

Washers. Washers are small metallic discs with holes in their centers. They are placed between the nut and one of the parts of a non-permanent connection. Washers serve to distribute tightening pressure over an area exceeding that of the nut, and they secure the connection. Two of the most common types of washers are the *plain flat washer* and the *spring lockwasher* (Figure 6-12). Both are available in standard dimensions.

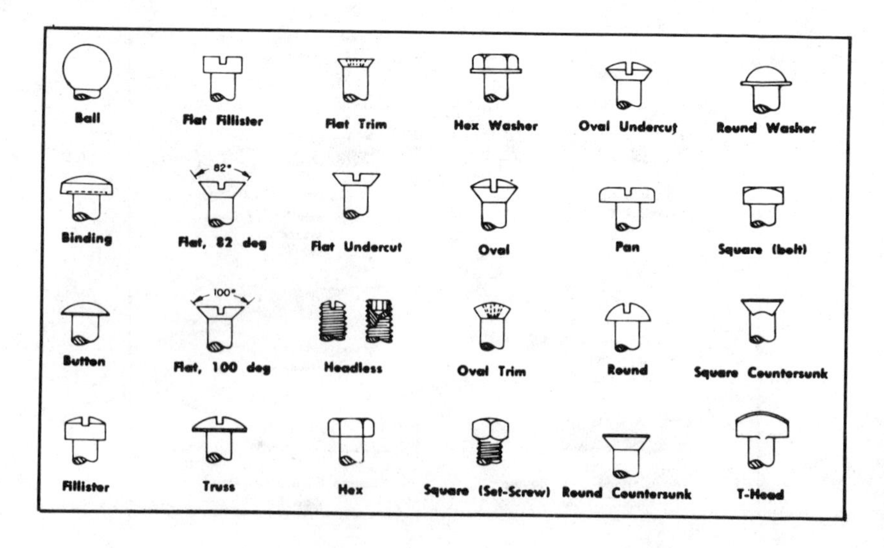

Figure 6-8. Typical standard bolt and screw head styles. (Machine Design)

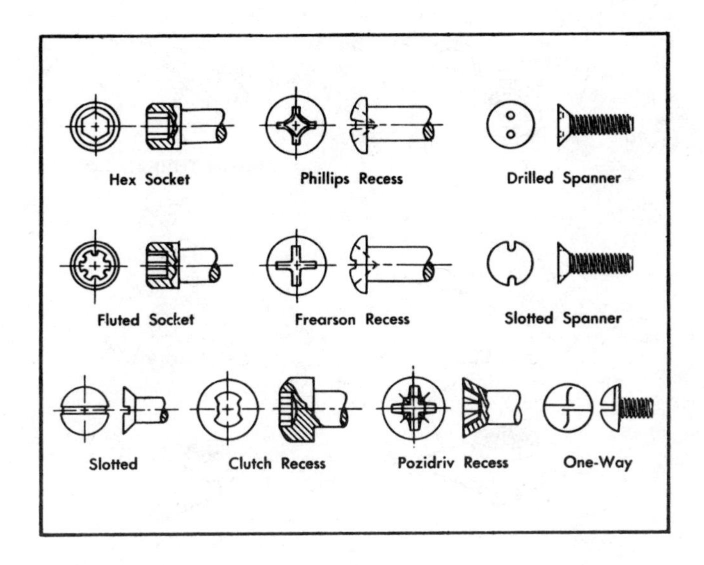

Figure 6-9. The various types of driving recesses in threaded fasteners. (Machine Design)

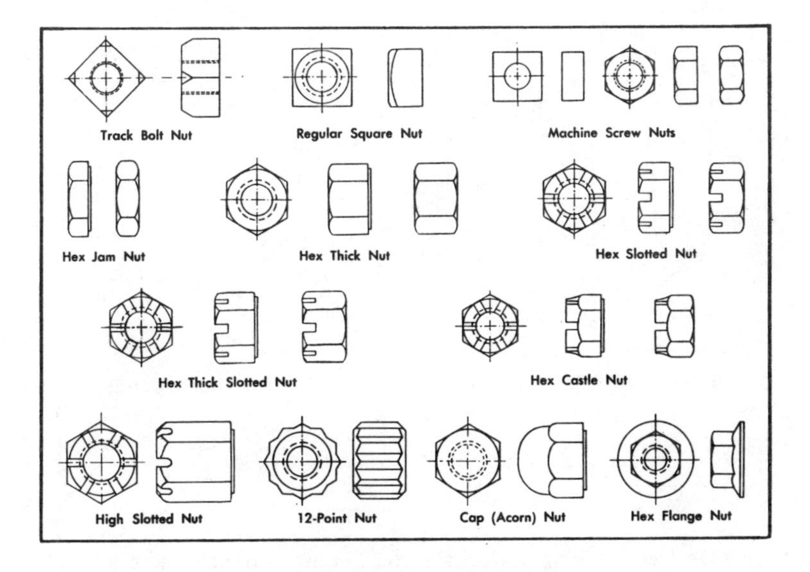

Figure 6-10. Standard nut styles. (Machine Design)

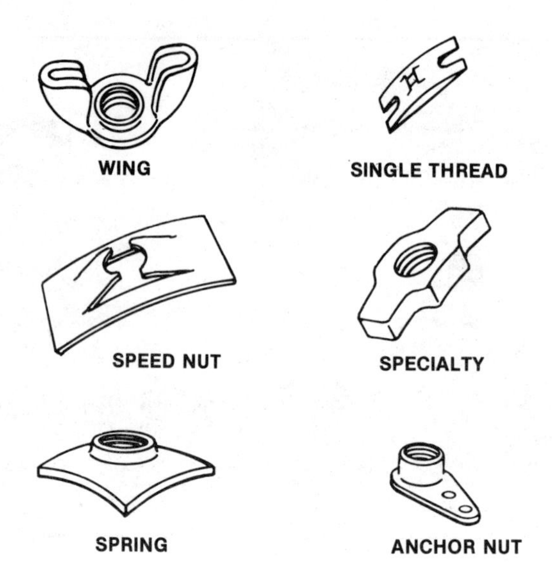

Figure 6-11. Typical specialty nut styles. (Deere and Company)

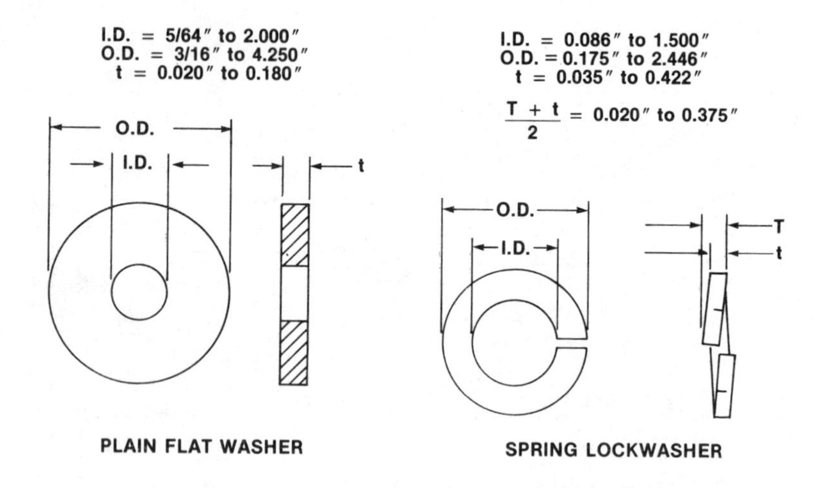

Figure 6-12. Plain flat washer and spring lockwasher.

Locking Devices. Locking devices are special nuts, washers, and pins used in non-permanent connections subjected to vibration. They prevent the nuts from loosening, and the parts from disengaging. Several types of locking devices exist (Figure 6-13).

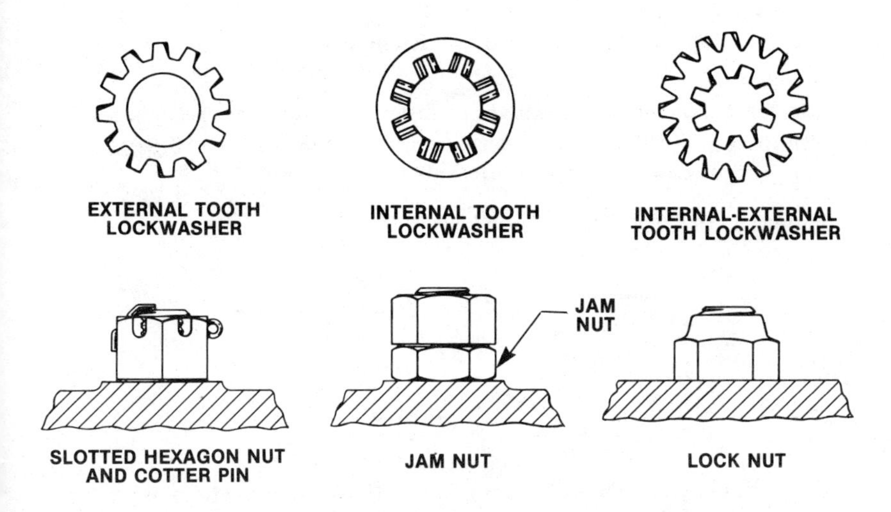

Figure 6-13. Typical locking devices.

MECHANICAL DRIVING SYSTEMS

Mechanical driving systems are mechanisms designed specifically for the transmission of power from one part of a machine to another part. Each mechanical driving system contains various machine elements. The most common machine elements used in mechanical driving systems are *shafts, couplings, keys, pins, V-belts, sheaves, roller chains, sprockets,* and *gears.*

Shafts

Shafts are cylindrical rods that are machined accurately to mate with bearings. They are made of medium-carbon or high-carbon steel. The shafts transfer power from one component of a machine to another component, or from one machine to another. The diameter of a shaft for a given application is determined by the strength of the steel from which it is made, the horsepower to be transmitted, the operating speed, and the torque. The following formulas are used to solve problems related to shafts:

$$\text{Torque (T)} = \frac{63{,}000\ \text{HP}}{\text{N}}$$

$$\text{Diameter (D)} = \sqrt[3]{\frac{5.1\text{T}}{\sigma_s}}$$

$$\text{Horsepower (HP)} = \frac{\text{D}^3\text{N}}{80}$$

HP = horsepower in ft lb
N = operating speed in RPM
T = torque in in lb
D = diameter of shaft in inches
σ_s = shearing strength of shaft

NOTE: 1 HP = 33,000 ft lb or 33,000(12) = 396,000 in lb. Also, shaft diameter can be determined by using the formula in Chapter 5, under TORSIONAL STRESS.

For applications of these formulas, see Example 6-2.

Example 6-2: Find the torque and the diameter of a shaft, then check the result.
Given: horsepower HP = 20, revolutions per minute N = 180 RPM
shearing stress σ_s = 4,000 lbs/in²

Solution: $$T = \frac{63{,}000HP}{N} = \frac{63{,}000(20)}{180} = 7{,}000 \text{ ft lb}$$

$$D = \sqrt[3]{\frac{5.1T}{\sigma_s}} = \sqrt[3]{\frac{5.1(7{,}000)}{4{,}000}} = \sqrt[3]{8.925} = 2.074''$$

Check: $$HP = \frac{D^3N}{80} = \frac{8.925(180)}{80} = 20.08 \text{ HP}$$

The result checks—the difference of 0.08 HP is due to the rounding off of the decimals in calculations.

Shafts are made in standard diameters to mate with the diameters of bores in bearings, couplings, sheaves, sprocket wheels, and gears. These diameters range in size from $\frac{1}{2}''$ to $3\frac{1}{4}''$ in increments of $\frac{1}{16}''$ and $\frac{1}{8}''$.

Couplings

Couplings are cast iron or steel components that have standard diameters similar to those of shafts. They are used to connect shafts that lie on the same axis line, or slightly off-center of the axis line. Couplings are classified as either *rigid* or *flexible* (Figure 6-14).

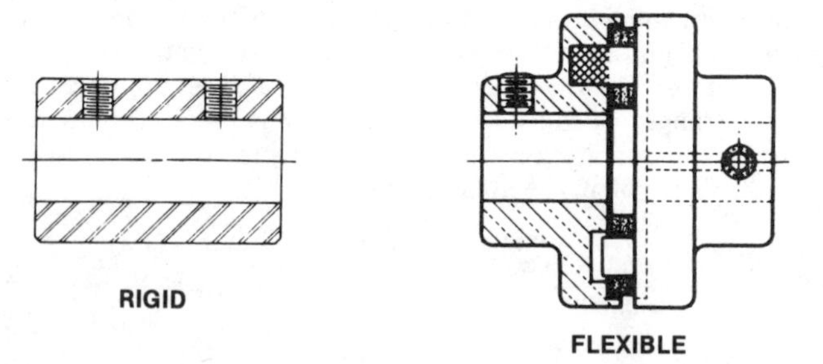

Figure 6-14. Sectional views of rigid and flexible couplings. (Browning Manufacturing Division)

Rigid couplings are used exclusively for connecting shafts that are in perfect, or almost perfect, alignment. Flexible couplings are used mainly for connecting shafts that are not in alignment.

Keys

Keys are small pieces of steel that have standard shapes and dimensions. They are used to prevent the free rotation of a pulley or gear on a shaft. The key is placed partially in the shaft keyseat and partially in the pulley or gear keyway (Figure 6-15).

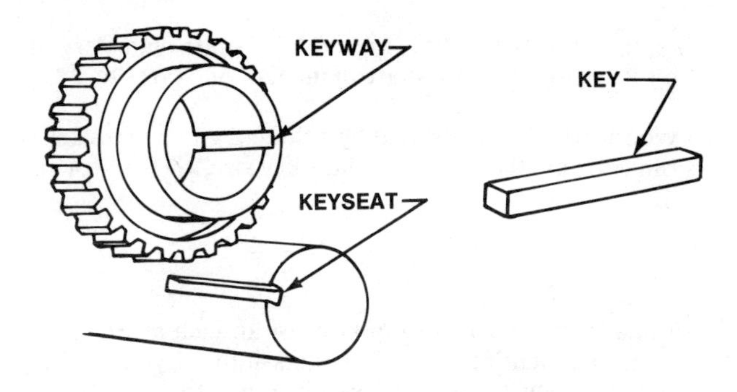

Figure 6-15. Key nomenclature. (Deere and Company)

The five types of keys are the *square, flat, gib-head (square and flat), Pratt and Whitney (sunk),* and *Woodruff* (Figure 6-16). As a rule, the part of the key that goes into the pulley or gear is equal to one half its width. The only exception to this rule is for Pratt and Whitney keys. They must have the equivalent of two thirds of their width engaging the shaft.

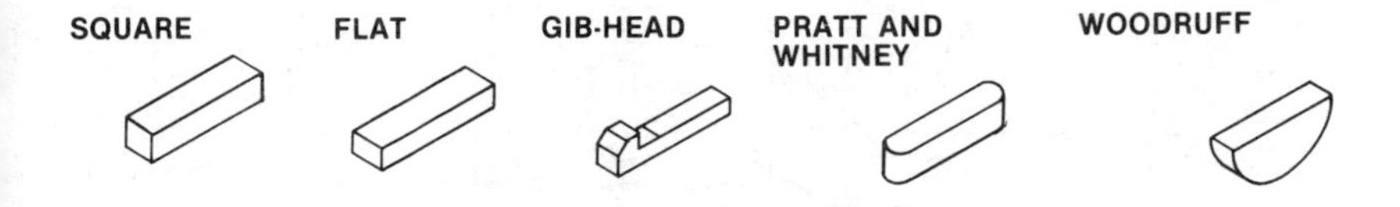

Figure 6-16. Types of keys.

Key dimensions have been standardized so that each key is suitable for a shaft that has a certain diameter:

Square and flat keys: From $\frac{1}{8}$" square or $\frac{1}{8}$" by $\frac{3}{32}$." for shaft diameters of $\frac{1}{2}$" to $\frac{9}{16}$" to $\frac{5}{8}$" square or $\frac{5}{8}$" by $\frac{7}{16}$" for shaft diameters of $2\frac{5}{16}$" to $2\frac{3}{4}$".

Gib-head keys: From $\frac{1}{8}$" square or $\frac{1}{8}$" by $\frac{3}{32}$" for shaft diameters of $\frac{1}{2}$" to $\frac{9}{16}$" to $1\frac{1}{2}$" square to $1\frac{1}{2}$" to 1" for shaft diameters of $5\frac{3}{4}$" to 6".

NOTE: Gib-head keys are tapered keys with a slope of $\frac{1}{8}$" per foot. The dimensions of their head (length and height) range from $\frac{1}{8}$" by $\frac{3}{16}$" to $1\frac{1}{2}$" by $1\frac{3}{4}$". They are suitable for large diameter shafts. The lengths of gib-head keys range from 4 to 16 times their width.

Pratt and Whitney keys: From 1/16″ by 3/32″ by 1/2″ for shaft diameters of 5/16″ to 3/8″ to 5/8″ by 15/16″ by 3″ for shaft diameters of 2 7/8″ to 3 3/4″.

NOTE: Each Pratt and Whitney key is identified by a single-digit or double-digit number, or a letter. The numbers range from 1 to 34, and the letters cover A to G. The smallest key size is No. 1, and the largest key size is No. 34.

Woodruff keys: From 1/16″ by 13/64″ by 1/2″ for shaft diameters of 5/16″ to 3/8″ to 3/8″ by 41/64″ by 1 1/2″ for shaft diameters of 1 13/16″ to 2″.

NOTE: Each Woodruff key is identified by a three-digit or four-digit number. The numbers range from 204 for the smallest key size, to 1012 for the largest key size.

Taper Pins

Taper pins are small pins that have a quarter of an inch taper-per-foot (Tpf = ¼″), and standard dimensions. The diameters at the large end of taper pins range from 0.0625″ to 0.4920″, and the lengths range from 0.375″ to 4.500″. Taper pins are used in shafts for transmitting very small amounts of torque and for positioning parts (Figure 6-17).

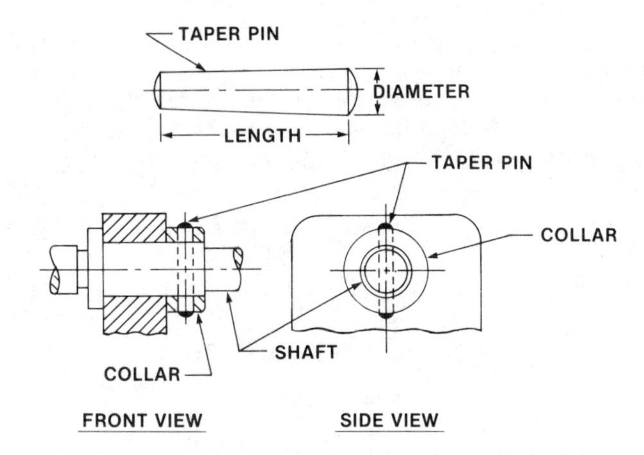

Figure 6-17. The taper pin and its use.

Cotter Pins

Cotter pins are small split pins that have standard dimensions (Figure 6-18). The nominal size of a cotter pin, identified by diameter A, ranges from 0.031″ to 0.725″, while the outside diameter of the eye, identified as diameter B, ranges from 1/16″ to 1 1/2″. The length of cotter pins ranges from ¾″ to 2″. Cotter pins secure assembled parts and prevent easy disassembly. For instance, a cotter pin can secure a slotted nut on the end of a threaded shaft or fastener as shown in Figure 6-18.

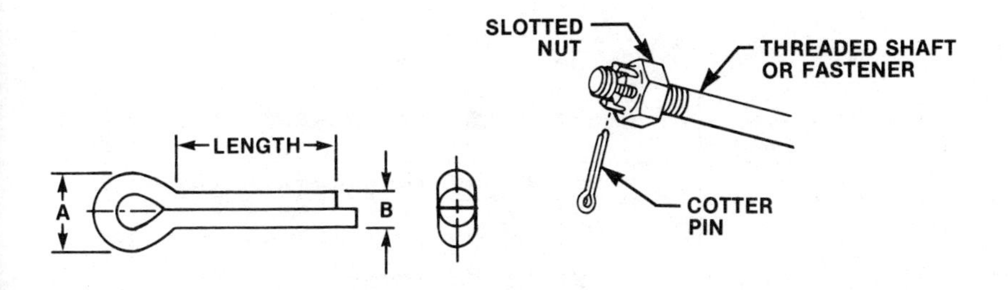

Figure 6-18. Cotter pin and its use. (Deere and Company)

V-Belts

V-belts are closed-looped belts made of rubber, or synthetic materials such as nylon, polyester, and rayon. They have a trapezoidal-shaped cross section and are used with grooved pulleys (sheaves) to transmit power from the drive to the driven shaft (Figure 6-19). The three types of V-belts are the *light-duty* or *fractional horsepower; standard;* and *heavy-duty narrow belts.*

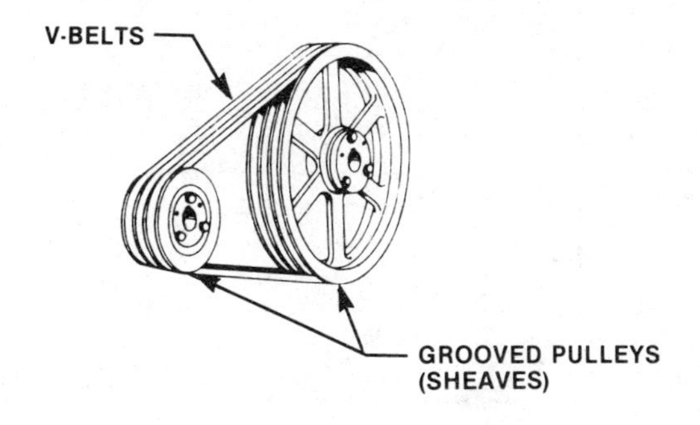

Figure 6-19. Typical V-belt application. (Browning Manufacturing Division)

Light-Duty or Fractional Horsepower V-Belts. Light-duty or fractional horsepower V-belts are manufactured in four standard sizes designated by the symbols 2L, 3L, 4L, and 5L (Figure 6-20). Each of these standard sizes has a specific width and thickness, and is available in standard lengths. Light-duty V-belts are designed for single-groove service on light-duty machinery drives such as refrigerators, pumps, heating and ventilating units, and bench equipment (machinery).

Standard V-Belts. Standard V-belts are manufactured in five standard sizes designated by the letters A, B, C, D, and E (Figure 6-21). Each of these standard sizes has a specific width and thickness, and is available in standard lengths. Standard V-belts are designed for heavy-duty service on machinery such as those used in the construction industry. The transfer of power is obtained by using two or more standard V-belts with the appropriate sheaves.

2L 3L 4L 5L

Belt No.	Length Outside	Length Pitch	Wt. Lbs.	Belt No.	Length Outside	Length Pitch	Wt. Lbs.	Belt No.	Length Outside	Length Pitch	Wt. Lbs.	Belt No.	Length Outside	Length Pitch	Wt. Lbs.
				3L540	54"	53.3"	.17	4L630	63"	62.0"	.31	5L420	42"	40.8"	.38
2L120	12	11.6	.03	3L550	55	54.3	.18	4L640	64	63.0	.38	5L430	43	41.8	.38
2L140	14	13.6	.04	3L560	56	55.3	.18	4L650	65	64.0	.38	5L440	44	42.8	.38
2L150	15	14.6	.04	3L570	57	56.3	.18	4L660	66	65.0	.38	5L450	45	43.8	.38
2L160	16	15.6	.04	3L580	58	57.3	.18	4L670	67	66.0	.38	5L460	46	44.8	.44
2L180	18	17.6	.05	3L590	59	58.3	.19	4L680	68	67.0	.38	5L470	47	45.8	.44
2L200	20	19.6	.06	3L600	60	59.3	.19	4L690	69	68.0	.38	5L480	48	46.8	.44
2L220	22	21.6	.06	3L610	61	60.3	.19	4L700	70	69.0	.38	5L490	49	47.8	.50
2L240	24	23.6	.07	3L620	62	61.3	.19	4L710	71	70.0	.38	5L500	50	48.8	.50
2L285	28½	28.1	.07	3L630	63	62.3	.20	4L720	72	71.0	.38	5L510	51	49.8	.50
2L300	30	29.6	.08	4L170	17	16.0	.10	4L730	73	72.0	.38	5L520	52	50.8	.50
2L310	31	30.6	.08	4L180	18	17.0	.10	4L740	74	73.0	.38	5L530	53	51.8	.50
2L320	32	31.6	.09	4L190	19	18.0	.11	4L750	75	74.0	.44	5L540	54	52.8	.50
2L325	32½	32.1	.09	4L200	20	19.0	.11	4L760	76	75.0	.44	5L550	55	53.8	.50
2L345	34½	34.1	.09	4L210	21	20.0	.12	4L770	77	76.0	.44	5L560	56	54.8	.50
				4L220	22	21.0	.12	4L780	78	77.0	.44	5L570	57	55.8	.50
3L110	11	10.3	.03	4L225	22½	21.5	.13	4L790	79	78.0	.44	5L580	58	56.8	.50
3L120	12	11.3	.04	4L230	23	22.0	.13	4L800	80	79.0	.44	5L590	59	57.8	.50
3L130	13	12.3	.04	4L240	24	23.0	.13	4L810	81	80.0	.44	5L600	60	58.8	.56
3L140	14	13.3	.05	4L245	24½	23.5	.13	4L820	82	81.0	.44	5L610	61	59.8	.56
3L150	15	14.3	.05	4L250	25	24.0	.13	4L830	83	82.0	.44	5L620	62	60.8	.56
3L160	16	15.3	.05	4L260	26	25.0	.13	4L840	84	83.0	.44	5L630	63	61.8	.56
3L170	17	16.3	.05	4L270	27	26.0	.13	4L850	85	84.0	.50	5L640	64	62.8	.63
3L180	18	17.3	.06	4L280	28	27.0	.13	4L860	86	85.0	.50	5L650	65	63.8	.63
3L190	19	18.3	.06	4L290	29	28.0	.13	4L870	87	86.0	.50	5L660	66	64.8	.63

(Browning Manufacturing Division)

Figure 6-20. Typical light-duty or fractional horsepower V-belt symbols, dimensions, and weights.

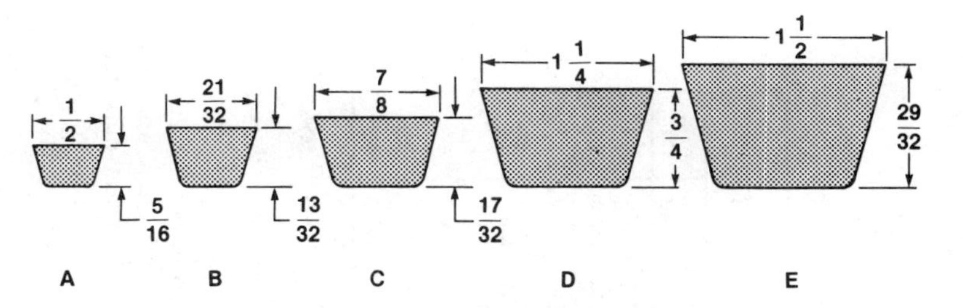

Belt No.	Length Outside	Length Pitch	Wt. Lbs.	Belt No.	Length Outside	Length Pitch	Wt. Lbs.	Belt No.	Length Outside	Length Pitch	Wt. Lbs.	Belt No.	Length Outside	Length Pitch	Wt. Lbs.
A35	37.2"	36.3"	.2	B36	39.0"	37.8"	.4	C51	55.2"	53.9"	1.0	D330	332.7"	330.8"	10.6
A36	38.2	37.3	.3	B38	41.0	39.8	.4	C60	64.2	62.9	1.2	D345	347.7	345.8	11.0
A37	39.2	38.3	.3	B40	43.0	41.8	.5	C68	72.2	70.9	1.3	D360	362.7	360.8	11.5
A38	40.2	39.3	.3	B41	44.0	42.8	.5	C72	76.2	74.9	1.4	D390	392.7	390.8	12.4
A39	41.2	40.3	.3	B42	45.0	43.8	.5	C75	79.2	77.9	1.4	D420	422.7	420.8	13.4
A40	42.2	41.3	.3	B43	46.0	44.8	.5	C81	85.2	83.9	1.6	E195	202.0	199.5	10.0
A41	43.2	42.3	.3	B44	47.0	45.8	.5	C85	89.2	87.9	1.6	E210	217.0	214.5	10.8
A42	44.2	43.3	.3	B45	48.0	46.8	.5	C90	94.2	92.9	1.7	E225	228.5	226.0	11.6
A43	45.2	44.3	.3	B46	49.0	47.8	.5	C96	100.2	98.9	1.8	E240	243.5	241.0	12.3
A44	46.2	45.3	.3	B47	50.0	48.8	.5	C100	104.2	102.9	1.9	E270	273.5	271.0	13.9
A45	47.2	46.3	.3	B48	51.0	49.8	.5	C105	109.2	107.9	2.0	E300	303.5	301.0	15.4
A46	48.2	47.3	.3	B49	52.0	50.8	.6	C109	113.2	111.9	2.0	E330	333.5	331.0	17.0
A47	49.2	48.3	.3	B50	53.0	51.8	.6	C112	116.2	114.9	2.1	E360	363.5	361.0	18.5
A48	50.2	49.3	.3	B51	54.0	52.8	.6	D255	257.7	255.8	8.1	E390	393.5	391.0	20.0
A49	51.2	50.3	.4	B52	55.0	53.8	.6	D270	272.7	270.8	8.9	E420	423.5	421.0	21.6
A90	92.2	91.3	.6	B53	56.0	54.8	.6	D285	287.7	285.8	9.8	E480	483.5	481.0	24.7
A91	93.2	92.3	.6	B54	57.0	55.8	.6	D300	302.7	300.8	10.5	E540	543.5	541.0	27.8
A92	94.2	93.3	.6	B55	58.0	56.8	.6	D315	317.7	315.8	10.2	E600	603.5	601.0	30.8
A93	95.2	94.3	.6	B55	58.0	56.8	.6					E660	663.5	661.0	34.0
A94	96.2	95.3	.6	B56	59.0	57.8	.6								

(Browning Manufacturing Division)

Figure 6-21. Typical standard V-belt symbols, dimensions, and weights.

Heavy-Duty Narrow V-Belts. Heavy-duty narrow V-belts are manufactured in three standard sizes designated by the symbols 3V, 5V, and 8V (Figure 6-22). Each of these standard sizes has a specific width and thickness and is available in standard lengths.

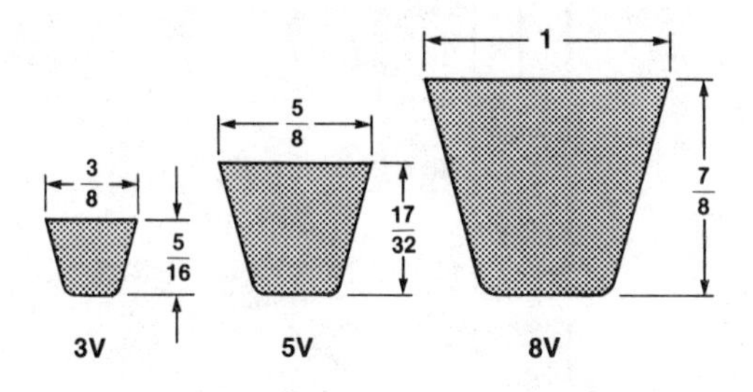

3V			5V			8V		
3/8" x 5/16"			5/8" x 17/32"			1" x 7/8"		
Part Number	Outside Length	Wt. Lbs.	Part Number	Outside Length	Wt. Lbs.	Part Number	Outside Length	Wt. Lbs.
3V250	25.0"	.1	5V500	50.0"	.6	8V1000	100.0"	3.3
3V265	26.5	.1	5V530	53.0	.6	8V1060	106.0	3.5
3V280	28.0	.1	5V560	56.0	.6	8V1120	112.0	3.6
3V300	30.0	.1	5V600	60.0	.7	8V1180	118.0	3.8
3V315	31.5	.1	5V630	63.0	.7	8V1250	125.0	3.9
3V335	33.5	.1	5V670	67.0	.8	8V1320	132.0	4.3
3V355	35.5	.2	5V710	71.0	.8	8V1400	140.0	4.5
3V375	37.5	.2	5V750	75.0	.8	8V1500	150.0	4.8
3V400	40.0	.2	5V800	80.0	.9	8V1600	160.0	5.1
3V425	42.5	.2	5V850	85.0	.9	8V1700	170.0	5.6
3V450	45.0	.2	5V900	90.0	1.0	8V1800	180.0	6.0
3V475	47.5	.2	5V950	95.0	1.0	8V1900	190.0	6.3
3V500	50.0	.2	5V1000	100.0	1.1	8V2000	200.0	6.5

(Browning Manufacturing Division)

Figure 6-22. Typical heavy-duty narrow V-belt symbols, dimensions, and weights.

Heavy-duty narrow V-belts require a smaller cross section per horsepower than the standard V-belt, and smaller sized sheaves; thus, the weight and cost of the V-belt drives are reduced. Like the standard V-belts, the heavy-duty narrow V-belts are also used on heavy-duty machinery. Also, the transfer of power is obtained by using two or more V-belts with the appropriate sheaves.

Sheaves

Sheaves are grooved pulleys that conform to the cross section of V-belts. They are usually made of aluminum, cast iron, or formed-steel. The three types of sheaves are *spoked-cast sheaves* with integral hubs, *disk-type cast sheaves* with removable hubs, and *formed-steel, light-duty sheaves* with integral hubs (Figure 6-23).

The shape and size of sheave grooves have been standardized. Each V-belt can only be used with sheaves that have a certain minimum pitch diameter. The groove dimensions of sheaves used with standard V-belts, along with a formula for finding face width (FW), are shown in Figure 6-24.

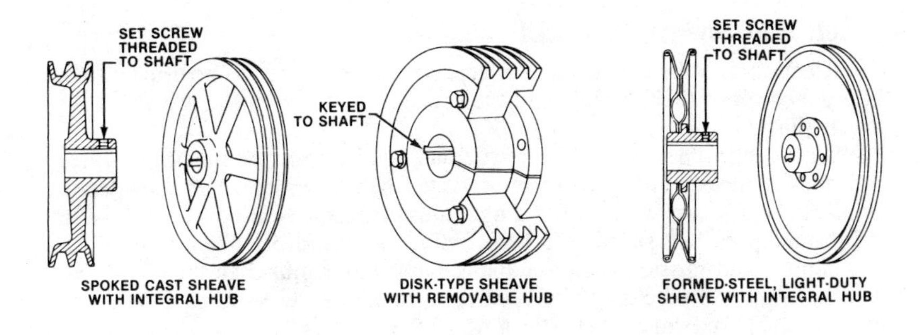

Figure 6-23. Types of sheaves for V-belt drives. (Deere and Company)

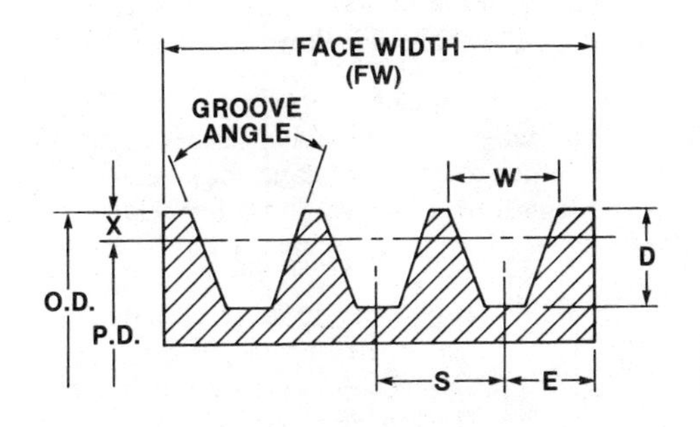

FW = S(N − 1) + 2E
N = NUMBER OF GROOVES

STANDARD V-BELT	MINIMUM RECOMMENDED PITCH DIAM.	PITCH DIAMETER P.D.	GROOVE ANGLE	W	D	X	S	E
A	3.0	2.6 to 5.4 Over 5.4	34° 38°	.494 .504	.490	.125	$^5/_8$	$^3/_8$
B	5.4	4.6 to 7.0 Over 7.0	34° 38°	.637 .650	.580	.175	$^3/_4$	$^1/_2$
C	9.0	7.0 to 7.99 8.0 to 12.0 Over 12.0	34° 36° 38°	.879 .887 .895	.780	.200	1	$^{11}/_{16}$
D	13.0	12.0 to 12.99 13.0 to 17.0 Over 17.0	34° 36° 38°	1.259 1.271 1.283	1.050	.300	$1^7/_{16}$	$^7/_8$
E	21.0	18.0 to 24.0 Over 24.0	36° 38°	1.527 1.542	1.300	.400	$1^3/_4$	$1^1/_8$

Figure 6-24. Standard sheave groove dimensions and face width formula.

V-Belt and Sheave Selection

The selection of V-belts and sheaves depends on the power to be transmitted, the surface speed (belt speed), and the speed ratio between the drive and the driven shaft.

When selecting a V-belt for a particular application, the maximum-rated transmission power and the maximum-rated belt speed must be considered. Light-duty belts can be used for a maximum transmission power of 7.5 HP and a maximum belt speed of 5,000 ft/min. Standard belts can be used for a maximum transmission power of 300 HP and a maximum belt speed of 6,000 ft/min. Heavy-duty belts can be used for a maximum transmission power of 500 HP and a maximum belt speed of 10,000 ft/min.

When selecting sheaves, the minimum diameter of the small sheave and the material of the sheave (composition) must be considered. The minimum sheave diameter must not be smaller than the manufacturer's recommended diameter. The sheave material must be suitable for the maximum speed of the belt drive.

NOTE: Manufacturers of transmission equipment typically provide tables with recommendations regarding the proper combination of V-belts and sheaves for specific service requirements. It is always advisable to follow the manufacturer's recommendations.

In every V-belt driving system, the length of the V-belt depends on the diameters of the two sheaves and the center-to-center distance between the two shafts. The formulas for finding the length of a V-belt, and other related formulas are:

$$L = 2C + 1.57(D + d) + \frac{(D - d)^2}{4C}$$

$$C = \frac{b + \sqrt{b^2 - 32(D - d)^2}}{16}$$

$$b = 4L - 6.28(D - d)$$

L = outside or pitch length of belt in inches
C = center distance in inches
D = outside or pitch diameter of large sheave in inches
d = outside or pitch diameter of small sheave in inches
b = constant

For an application of these formulas, see Examples 6-3 and 6-4.

Example 6-3: Find the pitch length of a standard V-belt size.

Given: Pitch diameter of small sheave d = 1″
Pitch diameter of large sheave D = 3″
Center distance between the shafts C = 15″

Solution:
$$L = 2C + 1.57(D + d) + \frac{(D - d)^2}{4C}$$

$$= 2(15) + 1.57(3 + 1) + \frac{(3 - 1)^2}{4(15)}$$

$$= 30 + 6.28 + .067 = 36.35''$$

NOTE: The closest belt number for this length, as shown in Figure 6-21, is A35.

Example 6-4: Find the center distance between the two shafts of a V-belt drive that uses a standard A90 V-belt.

Given: Pitch length of A90 V-belt = 91.3″ (obtained from Figure 6-21)
Pitch diameter of small sheave d = 3″
Pitch diameter of large sheave D = 12″

Solution: $b = 4L - 6.28(D + d)$

$= 4(91.3) - 6.28(15) = 365.2 - 94.2 = 271$

$$C = \frac{b + \sqrt{b^2 - 32(D - d)^2}}{16} = \frac{271 + \sqrt{271^2 - 32(12 - 3)^2}}{16}$$

$$= \frac{271 + \sqrt{70{,}849}}{16} = \frac{271 + 266.18}{16} = 33.574''$$

Roller Chains

Roller chains, utilizing sprockets, are used to transmit power in parallel shafts. Roller chains consist of a series of alternate pin linkplates, roller linkplates, rollers, pins, and bushings (Figure 6-25). In every roller-pin-bushing joint, the pin is free to pivot within the bushing, and the roller is free to turn on the bushing. If the roller chain is detachable, as is the one shown in Figure 6-25, cotter pins are used in the pin holes.

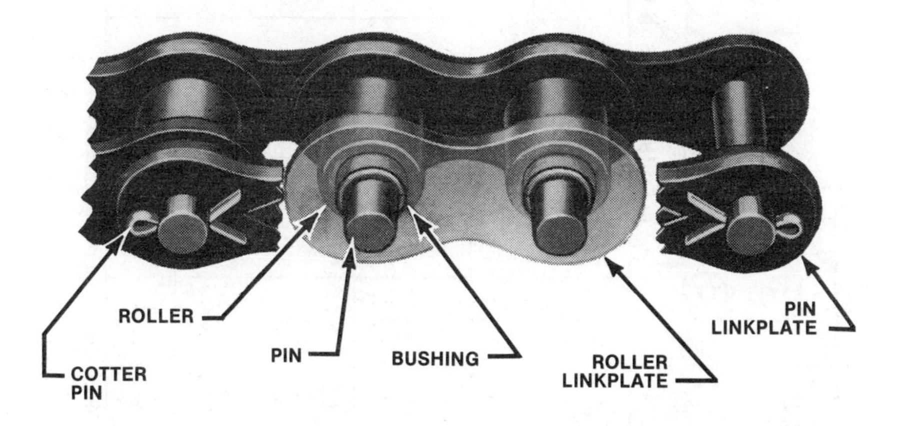

Figure 6-25. Component parts of a typical roller chain. (Diamond Chain Company)

Several types of roller chains designed for various applications exist. The most common types are the *standard series chain, heavy series chain, double-pitch drive chain,* and *double-pitch conveyor chain.*

Standard Series Chain. The standard series chain is considered the basic power transmission roller chain. It is available in ¼″ to 2½″ pitches in single or multiple strands. It conforms to both ANSI and ISO standards. Figure 6-26 shows a typical list of standard series chain sizes and related data.

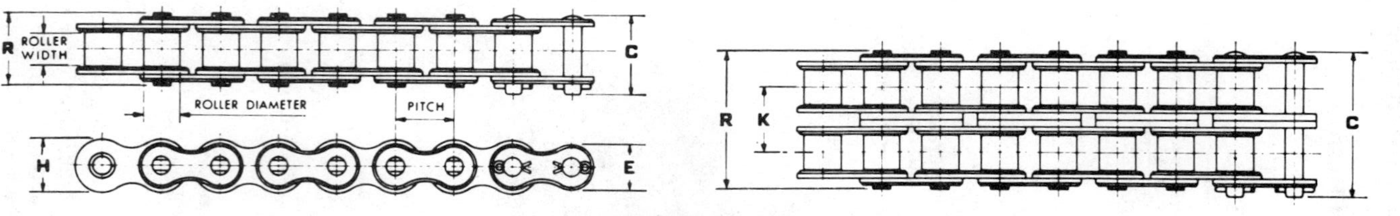

ALL DIMENSIONS IN INCHES

ANSI & Diamond Number	ISO Number	Pitch Inches	Roller Width	Roller Diam.	Pin Diam.	Linkplate			C	R	K	Bearing Area Sq. Inch	Weight Per Foot Pounds	Average Tensile Strength
						Thick-ness	E Height	H Height						
STANDARD SERIES CHAINS														
25	04C-1	¼	⅛	*.130	.090	.030	.205	.237	.37	.34	....	.017	.084	875
25-2	04C-2	¼	⅛	*.130	.090	.030	.205	.237	.63	.59	.252	.034	.163	1750
25-3	04C-3	¼	⅛	*.130	.090	.030	.205	.237	.88	.84	.252	.051	.246	2625
40	08A-1	½	5/16	.312	.156	.060	.410	.475	.72	.67	....	.067	.41	3700
40-2	08A-2	½	5/16	.312	.156	.060	.410	.475	1.29	1.24	.566	.134	.80	7400
40-3	08A-3	½	5/16	.312	.156	.060	.410	.475	1.85	1.80	.566	.201	1.20	11100
40-4	08A-4	½	5/16	.312	.156	.060	.410	.475	2.42	2.37	.566	.268	1.60	14800
40-6	08A-6	½	5/16	.312	.156	.060	.410	.475	3.56	3.51	.566	.402	2.42	22200
41	085	½	¼	.306	.141	.050	.310	.383	.65	.57	....	.049	.26	2000

*Chains are rollerless—dimension shown is bushing diameter.

Figure 6-26. Typical listing of standard series chains and related information. (Diamond Chain Company)

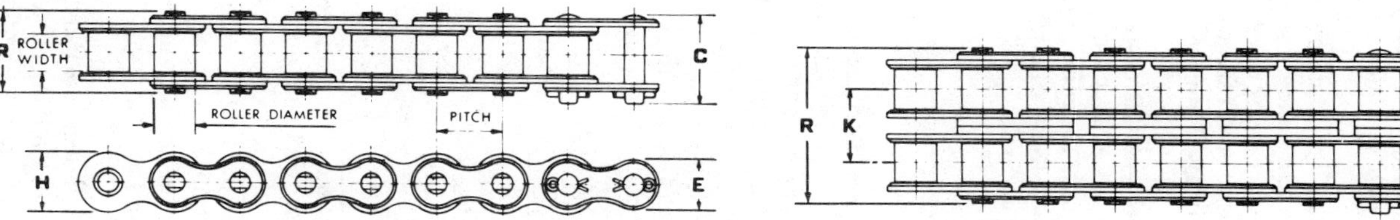

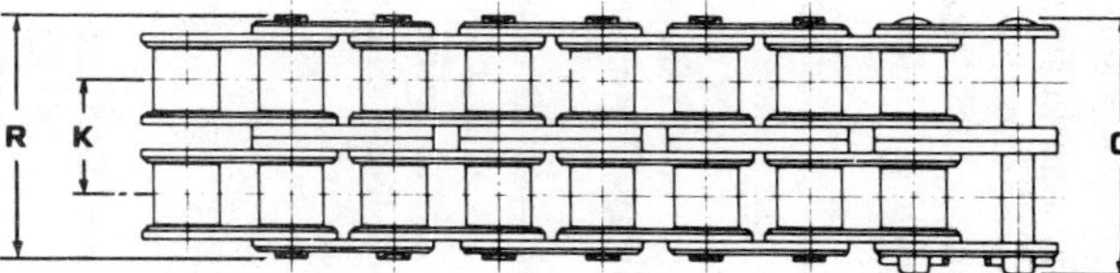

DIMENSIONS IN INCHES

ANSI & Diamond Number	ISO Number	Pitch Inches	Roller Width	Roller Diam.	Pin Diam.	Linkplate			C	R	K	Bearing Area Sq. Inch	Weight Per Foot Pounds	Average Tensile Strength
						Thick-ness	E Height	H Height						
HEAVY SERIES CHAINS														
60H	....	3/4	1/2	.469	.234	.125	.615	.712	1.24	1.17	....	.176	1.18	8500
60H-2	....	3/4	1/2	.469	.234	.125	.615	.712	2.27	2.20	1.028	.352	2.33	17000
60H-3	....	3/4	1/2	.469	.234	.125	.615	.712	3.31	3.24	1.028	.528	3.47	25500
60H-4	....	3/4	1/2	.469	.234	.125	.615	.712	4.34	4.26	1.028	.704	4.61	34000
80H	....	1	5/8	.625	.312	.156	.820	.950	1.57	1.45	....	.295	2.02	14500
80H-2	....	1	5/8	.625	.312	.156	.820	.950	2.84	2.72	1.283	.590	3.93	29000
80H-3	....	1	5/8	.625	.312	.156	.820	.950	4.14	4.02	1.283	.885	5.92	43500
80H-4	....	1	5/8	.625	.312	.156	.820	.950	5.42	5.30	1.283	1.180	7.87	58000

Figure 6-27. Typical listing of heavy series chain and related information. (Diamond Chain Company)

Heavy Series Chain. The heavy series chain is similar to the standard series chain except that the linkplate thickness for a given pitch is increased to the thickness corresponding to the next largest pitch. The increased linkplate thickness provides additional capacity to withstand a shock load (a suddenly applied load). The heavy series chain conforms to both ANSI and ISO standards and is available in ¾ ″ to 2½ ″ pitches in single and multiple strands. Figure 6-27 shows a typical listing of heavy series chain sizes and related data.

Double-Pitch Drive Chain. The double-pitch drive chain is a light-duty power transmission roller chain. Its dimensions are similar to those of a standard series chain except that the pitch is doubled. The linkplates have a contour resembling a "figure eight." The double pitch drive chain conforms to ANSI standards and is available in 1 ″ to 2 ″ pitches, but only in single strands (Figure 6-28).

Double-Pitch Conveyor Chain. The double-pitch conveyor chain is designed for conveyor service. It conforms to ANSI standards and has oval contoured linkplates. It is available in 1 ″ to 4 ″ pitches with single strands for either standard or large diameter rollers. Refer to Figure 6-28 for a typical listing of double-pitch drive chain and double-pitch conveyor chain sizes and related data.

Sprockets

Sprockets are toothed wheels designed for use with roller chains. They are usually made from cast iron and cast steel castings, or steel plates. The four types of sprockets (Figure 6-29) are: *Type A* - plain plate without hub; *Type B* - casting with a hub on one side; *Type C* - casting with hubs on both sides; *Type D* - plain plate attached to a hub.

Each sprocket is characterized by four diameters (Figure 6-30). One of these diameters is the pitch diameter, which is on a theoretical circle. Its size depends on the roller chain pitch and the number of teeth in the sprocket. Pitch diameter is found by using the formula,

$$\text{Pitch diameter} = \frac{\text{Pitch of roller chain}}{\text{sine}(180^\circ/\text{Number of teeth})}$$

The pitch diameter is of primary importance because all of the other diameters are based on it. For instance, the bottom diameter of a sprocket with an odd number of teeth (Figure 6-30) cannot be measured, but its caliber diameter can be measured. The caliber diameter is the dimension across the bottom of the tooth spaces which are most nearly opposite each other, but this does not represent the exact bottom diameter. The exact bottom diameter of an odd-toothed sprocket can only be found by subtracting the roller chain diameter from the pitch diameter.

Sprockets are made for single or multiple chain strands. Their teeth are usually cut by production tooling, but there are also specially designed milling cutters for standard roller chain sizes that can be used to cut sprockets. Typical dimensions for a No. 41 sprocket are shown in Figure 6-31.

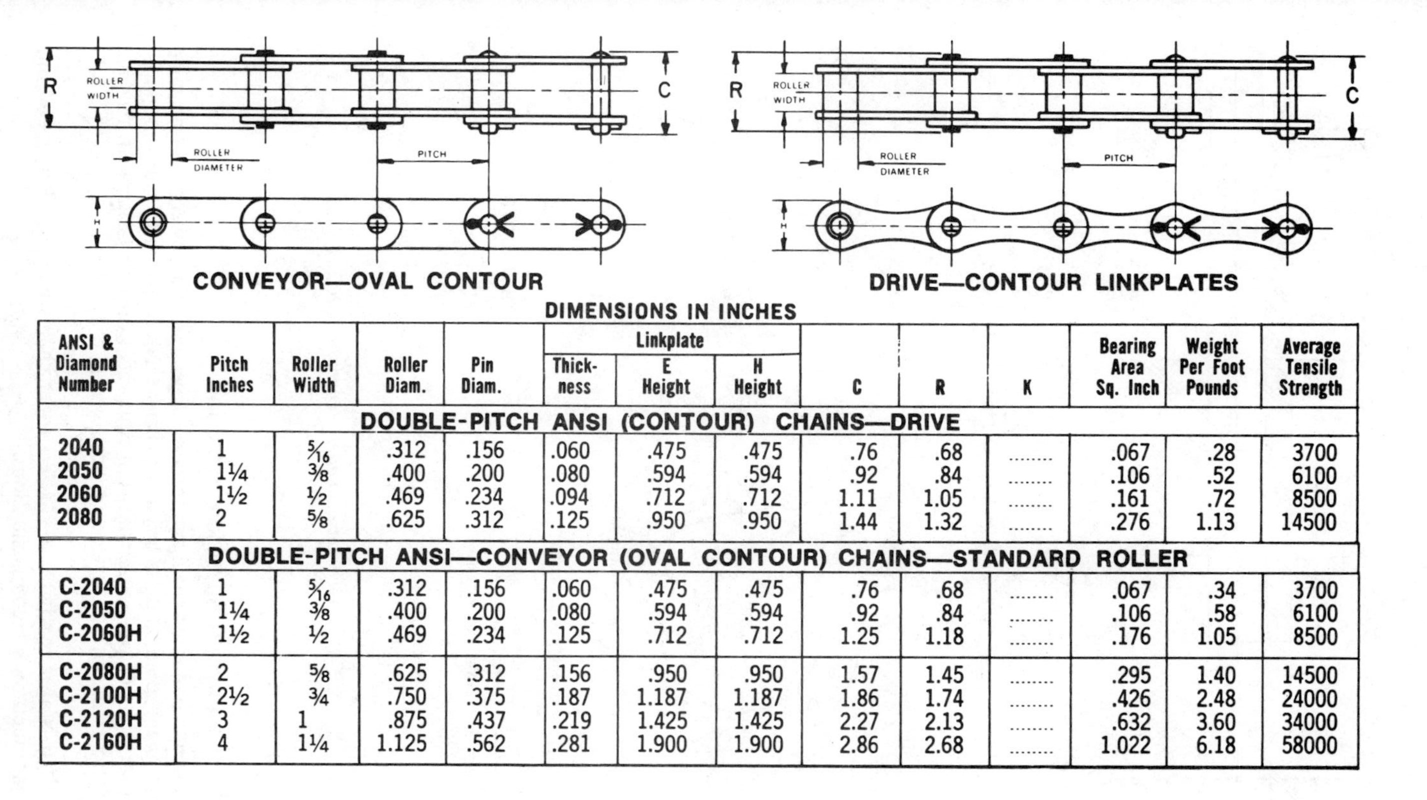

DIMENSIONS IN INCHES

ANSI & Diamond Number	Pitch Inches	Roller Width	Roller Diam.	Pin Diam.	Linkplate			C	R	K	Bearing Area Sq. Inch	Weight Per Foot Pounds	Average Tensile Strength
					Thick-ness	E Height	H Height						
DOUBLE-PITCH ANSI (CONTOUR) CHAINS—DRIVE													
2040	1	5/16	.312	.156	.060	.475	.475	.76	.68	....	.067	.28	3700
2050	1¼	3/8	.400	.200	.080	.594	.594	.92	.84	....	.106	.52	6100
2060	1½	½	.469	.234	.094	.712	.712	1.11	1.05	....	.161	.72	8500
2080	2	5/8	.625	.312	.125	.950	.950	1.44	1.32	....	.276	1.13	14500
DOUBLE-PITCH ANSI—CONVEYOR (OVAL CONTOUR) CHAINS—STANDARD ROLLER													
C-2040	1	5/16	.312	.156	.060	.475	.475	.76	.68	....	.067	.34	3700
C-2050	1¼	3/8	.400	.200	.080	.594	.594	.92	.84	....	.106	.58	6100
C-2060H	1½	½	.469	.234	.125	.712	.712	1.25	1.18	....	.176	1.05	8500
C-2080H	2	5/8	.625	.312	.156	.950	.950	1.57	1.45	....	.295	1.40	14500
C-2100H	2½	¾	.750	.375	.187	1.187	1.187	1.86	1.74	....	.426	2.48	24000
C-2120H	3	1	.875	.437	.219	1.425	1.425	2.27	2.13	....	.632	3.60	34000
C-2160H	4	1¼	1.125	.562	.281	1.900	1.900	2.86	2.68	....	1.022	6.18	58000

Figure 6-28. Typical listing of double-pitch drive chains, double-pitch conveyor chains, and related information. (Diamond Chain Company)

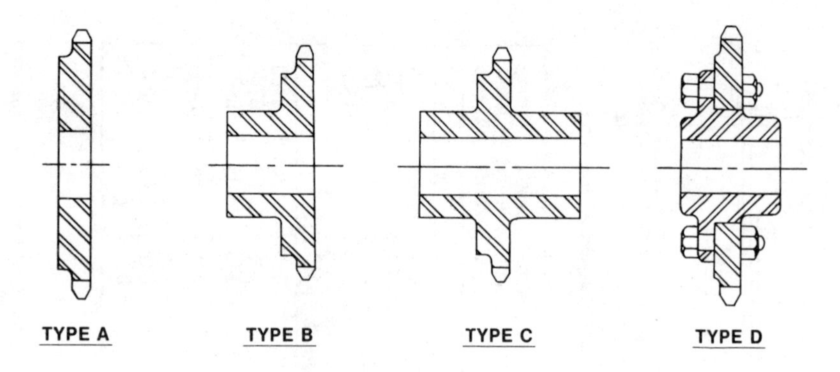

Figure 6-29. The four types of sprockets.

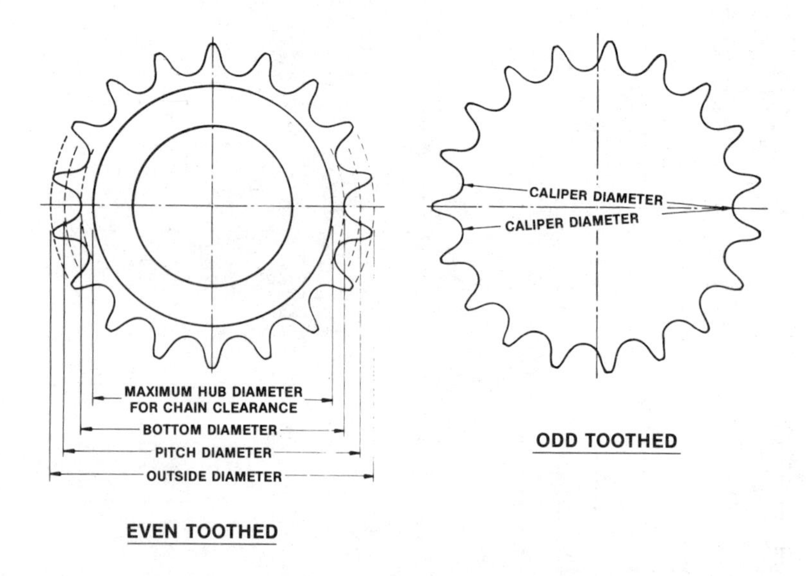

Figure 6-30. Sprocket diameters. (Diamond Chain Company)

Roller Chain and Sprocket Selection

The selection of a roller chain and sprocket depends on the power to be transmitted, the chain speed, and the speed ratio between the drive and the driven shaft. Manufacturers of power transmission equipment provide tables with recommendations for various service applications.

NOTE: It is always advisable to select roller chains and sprockets based on the manufacturer's recommendations.

No. of Teeth	Pitch Diameter	Outside Diameter	Bottom Diam. Even Teeth Caliper Diam. Odd Teeth	Max. Hub Diam.	Max. Bore Diam.	No. of Teeth	Pitch Diameter	Outside Diameter	Bottom Diam. Even Teeth Caliper Diam. Odd Teeth	No. of Teeth	Pitch Diameter	Outside Diameter	Bottom Diam. Even Teeth Caliper Diam. Odd Teeth
6	1.000	1.17	.694	.43	.12	**54**	8.599	8.89	8.293	**102**	16.236	16.53	15.930
7	1.152	1.34	.817	.60	.31	**55**	8.758	9.04	8.449	**103**	16.395	16.69	16.087
8	1.307	1.51	1.001	.77	.50	**56**	8.917	9.20	9.611	**104**	16.555	16.85	16.249
9	1.462	1.67	1.134	.94	.56	**57**	9.076	9.36	8.767	**105**	16.714	17.01	16.406
10	1.618	1.84	1.312	1.10	.69	**58**	9.236	9.52	8.930	**106**	16.873	17.17	16.567
11	1.775	2.00	1.451	1.27	.88	**59**	9.395	9.68	9.086	**107**	17.032	17.33	16.724
12	1.932	2.17	1.626	1.43	.88	**60**	9.554	9.84	9.248	**108**	17.191	17.48	16.885
13	2.089	2.33	1.768	1.59	1.06	**61**	9.713	10.00	9.404	**109**	17.350	17.64	17.042
14	2.247	2.49	1.941	1.75	1.25	**62**	9.872	10.16	9.566	**110**	17.509	17.80	17.203
15	2.405	2.65	2.086	1.92	1.25	**63**	10.031	10.32	9.722	**111**	17.669	17.96	17.361
16	2.563	2.81	2.257	2.08	1.31	**64**	10.190	10.48	9.884	**112**	17.828	18.12	17.522
17	2.721	2.98	2.403	2.24	1.44	**65**	10.349	10.64	10.040	**113**	17.987	18.28	17.679
18	2.879	3.14	2.573	2.40	1.62	**66**	10.508	10.80	10.202	**114**	18.146	18.44	17.840
19	3.038	3.30	2.722	2.56	1.75	**67**	10.667	10.96	10.358	**115**	18.305	18.60	17.997
20	3.196	3.46	2.890	2.72	1.75	**68**	10.826	11.12	10.520	**116**	18.464	18.76	18.158
21	3.355	3.62	3.040	2.88	1.88	**69**	10.986	11.27	10.676	**117**	18.624	18.92	18.315
22	3.513	3.78	3.207	3.04	2.00	**70**	11.145	11.43	10.839	**118**	18.783	19.08	18.477
23	3.672	3.94	3.357	3.20	2.19	**71**	11.304	11.59	10.995	**119**	18.942	19.24	18.634
24	3.831	4.10	3.525	3.36	2.25	**72**	11.463	11.75	11.157	**120**	19.101	19.39	18.795
25	3.989	4.26	3.675	3.52	2.25	**73**	11.622	11.91	11.313	**121**	19.260	19.55	18.952
26	4.148	4.42	3.842	3.68	2.38	**74**	11.781	12.07	11.475	**122**	19.419	19.71	19.113
27	4.307	4.58	3.994	3.84	2.56	**75**	11.940	12.23	11.631	**123**	19.578	19.87	19.270
28	4.466	4.74	4.160	4.00	2.75	**76**	12.099	12.39	11.793	**124**	19.737	20.03	19.431
29	4.625	4.90	4.312	4.16	2.75	**77**	12.258	12.55	11.949	**125**	19.897	20.19	19.589
30	4.783	5.06	4.477	4.32	2.81	**78**	12.417	12.71	12.111	**126**	20.056	20.35	19.750
31	4.942	5.22	4.630	4.48	2.94	**79**	12.577	12.87	12.268	**127**	20.215	20.51	19.907
32	5.101	5.38	4.795	4.64	3.12	**80**	12.736	13.03	12.430	**128**	20.374	20.67	20.068
33	5.260	5.54	4.948	4.80	3.18	**81**	12.895	13.19	12.587	**129**	20.533	20.83	20.226
34	5.419	5.70	5.113	4.96	3.25	**82**	13.054	13.34	12.748	**130**	20.692	20.99	20.386
35	5.578	5.86	5.266	5.12	3.31	**83**	13.213	13.50	12.905	**131**	20.851	21.15	20.544
36	5.737	6.02	5.431	5.28	3.50	**84**	13.372	13.66	13.066	**132**	21.011	21.31	20.705
37	5.896	6.18	5.585	5.44	3.62	**85**	13.531	13.82	13.223	**133**	21.170	21.46	20.862
38	6.055	6.33	5.749	5.60	3.75	**86**	13.690	13.98	13.384	**134**	21.329	21.62	21.023
39	6.214	6.49	5.903	5.26	3.81	**87**	13.849	14.14	13.541	**135**	21.488	21.78	21.180
40	6.373	6.65	6.067	5.92	3.88	**88**	14.008	14.30	13.702	**136**	21.647	21.94	21.341
41	6.532	6.81	6.221	6.08	4.00	**89**	14.168	14.46	13.860	**137**	21.806	22.10	21.499

Odd tooth bottom diameters equal pitch diameter minus .306″.
Maximum bore dimensions are for reference only.

Figure 6-31. Typical dimensions for a No. 41 sprocket used with a ½″ pitch standard series chain. (Diamond Chain Company)

In every roller-chain driving system, the length of the roller chain depends on the number of teeth of the drive and driven sprockets, and the center distance between the two shafts. The formula for finding the length of a roller chain, and a related formula are:

$$L = 2C + \frac{N}{2} + \frac{n}{2} + \left(\frac{N - n}{2\pi}\right)^2 \left(\frac{1}{C}\right)$$

$$C = \frac{P}{8}\left[2L - N - n + \sqrt{(2L - N - n)^2 - 0.810(N - n)^2}\right]$$

L = length of roller chain in chain pitches

NOTE: Pitch of a roller chain is the distance between the center holes of its linkplates in inches. The length of roller chains is expressed in chain pitches (number of pitches). For instance, the pitch of a roller chain is ½″; the length of 10 pitches of that chain would be 10 × ½″ = 5″.

C = center distance in inches
N = number of teeth of the large sprocket wheel (driven shaft)
n = number of teeth of the small sprocket wheel (drive shaft).

For applications of these formulas, see Examples 6-5 and 6-6.

Example 6-5: Find the length of a standard series roller chain.
Given: Center distance between the shafts C = 20″
Number of teeth of small sprocket n = 25
Number of teeth of large sprocket N = 45

Solution:
$$L = 2C + \frac{N}{2} + \frac{n}{2} + \left(\frac{N - n}{2\pi}\right)^2 \left(\frac{1}{C}\right)$$

$$= 2(20) + \frac{45}{2} + \frac{25}{2} + \left(\frac{45 - 25}{2(3.14)}\right)^2 \left(\frac{1}{20}\right)$$

$$= 40 + 35 + .507 = 75.507$$

$$L = 76 \text{ chain pitches}$$

Example 6-6: Find the center distance between the two shafts of a roller chain driving system that uses a No. 41 standard roller chain.

Given: Length of chain L = 120 pitches
Pitch P = ½″ or .5″ (obtained from Figure 6-26)
Number of teeth of the small sprocket wheel n = 17
Number of teeth of the large sprocket wheel N = 51

Solution: $$C = \frac{P}{8}\left[2L - N - n + \sqrt{(2L - N - n)^2 - 0.810(N - n)^2}\right]$$

$$= \frac{0.5}{8}\left[2(120) - 51 - 17 + \sqrt{(2 \times 120 - 51 - 17)^2 - 0.810(51 - 17)^2}\right]$$

$$= 0.0625\left[172 + \sqrt{29{,}584 - 936.36}\right]$$

$$= 0.0625(341.26) = 21.329''$$

Gears

Gears are toothed wheels with standard tooth forms and dimensions based on their circular pitch. They are used in pairs to transmit power when the drive and driven shafts are close to each other.

Gear names are based on their tooth form. The four most common types are the *spur gear, helical gear, bevel gear,* and *worm gear*. All of these gears can be made in conventional milling machines.

Spur Gears. Spur gears have straight teeth that are parallel to the shaft axes (Figure 6-32). These gears are used in applications where the shafts are parallel. They are suitable for speed ratios ranging from 1:1 to 1:6, and surface speeds of up to 1,000 feet per minute.

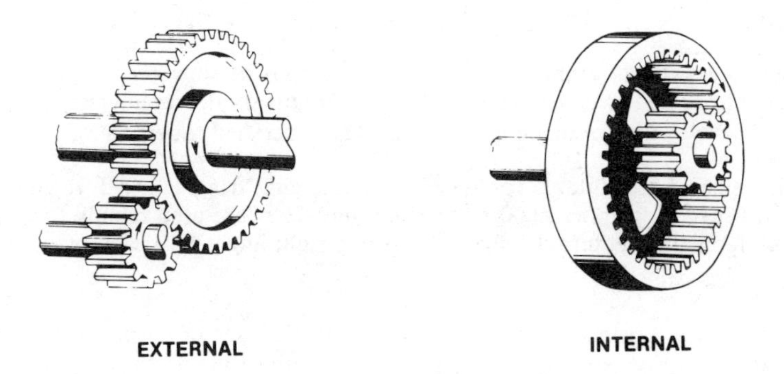

Figure 6-32. Types of spur gears. (Mobil Oil Corporation)

Helical Gears. Helical gears are similar to spur gears except their teeth are not parallel to the shaft axes, and they form a helix (Figure 6-33). They are made in pairs to match the helix axis of each other. Helical gears are used in gear driving systems with shafts that are parallel or skewed (not at right angles)—that is, they are non-parallel or non-intersecting.

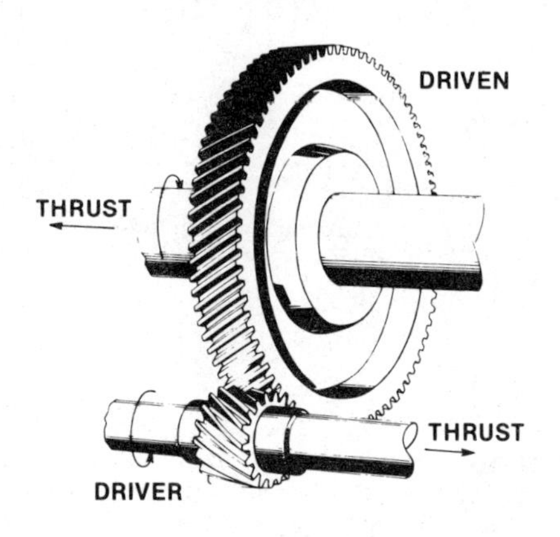

Figure 6-33. Helical gears appear to be similar to spur gears; however, their teeth are not parallel to the shaft axes. (Mobil Oil Corporation)

Helical gears are suitable for speed ratios ranging from 1:1 to 1:10, and for surface speeds of up to 5,000 feet per minute. The transmission of power via helical gears is smoother and less noisy than with spur gears.

Bevel Gears. Bevel gears are simple modifications of spur gears. They have tapered teeth and are used in applications where the shaft axes intersect (Figure 6-34). Bevel gears are made in pairs that have specified tapers to match each other.

Bevel gears are suitable for speed ratios ranging from 1:1 to 1:4, and for surface speeds of up to 1,000 feet per minute. They are used mainly for driving systems that require the shafts to be at a right angle with each other.

Worm Gears. Worm gears are made in pairs that consist of a *worm wheel* and a *worm gear* (Figure 6-35). In every pair, the teeth of the worm wheel are cut at an angle equal to the helix angle of the worm gear. These gears are used in gear driving systems in which the shafts are skewed.

Worm gears are suitable for speed ratios ranging from 1:3½ to 1:90, and for surface speeds of up to 4,000 feet per minute.

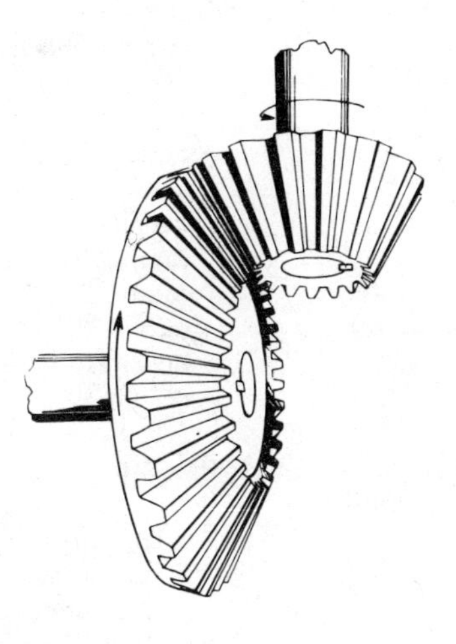

Figure 6-34. Bevel gears have tapered teeth and are used where shaft axes intersect. (Mobil Oil Corporation)

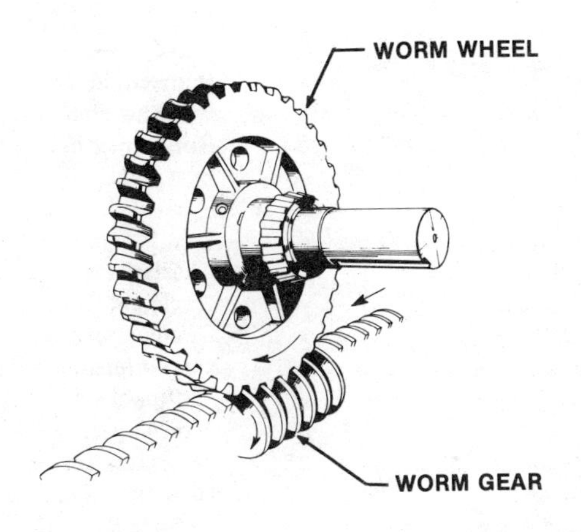

Figure 6-35. Worm gears consist of worm wheel and a worm gear. (Mobil Oil Corporation)

Index